Disease Suppressive Tropical Soils

Samuel Gnanamanickam • Monica Höfte
Editors

Disease Suppressive Tropical Soils

Springer

Editors
Samuel Gnanamanickam
R & D
Biotech Innovations (BTI AG LLC)
Wylie, TX, USA

Monica Höfte
Faculty of Bioscience Engineering
Ghent University
Gent, Belgium

ISBN 978-3-032-08099-8 ISBN 978-3-032-08100-1 (eBook)
https://doi.org/10.1007/978-3-032-08100-1

This Springer imprint is published by the registered company Springer Nature Switzerland AG
The registered company address is: Gewerbestrasse 11, 6330 Cham, Switzerland

Preface

The idea of developing a volume on Disease Suppressive Tropical Soils originated in Gnanamanickam's mind as a volume that will focus on Disease Suppressive Rice Soils in 2023. Hofte suggested that we extend the topic to Disease Suppressive Tropical Soils. By exchanging our ideas, we arrived at the title of the volume, and we felt also that extensive reviews about this topic did not exist. On the other hand, we realized also that Disease Suppressive Soils in the temperate environment have been exhaustively studied and written about in several outstanding publications.

Finding people with research expertise or experience on disease suppressive soils in the tropics was a challenge. We were fortunate to locate some of the most outstanding researchers who have researched important crops such as rice, banana, oil palm, or the kinds of stresses (salinity, drought, or other environmental stresses) which have been studied in the Tropics. The information that is assembled in this book is novel and should be of value to those seeking information in this area. Therefore, we believe that it serves as a valuable resource manual or a reference volume.

There is a need to study the tropical soils in depth. It is known that acid sulfate soils (ASS) are 24.0 million hectares worldwide. Asia has 6.5 million hectares (Mha); Africa has 4.5 Mha; Latin America has 3.0 Mha; Australia has 3.0 Mha; and North America has 100,000 ha. It is obvious that only few of these soils in the vast tropics have been carefully studied for the disease suppressive role they might play against economically important plant pathogens. Perhaps, at the present time when there is a push for excessive use of chemicals to achieve higher crop yields, this opens the way to find ecology-conscious biological alternatives/strategies that can preserve the soil and human health without compromising on crop yields!

Wylie, TX, USA — Samuel Gnanamanickam
Gent, Belgium — Monica Höfte

Acknowledgments

We are very grateful to all the chapter contributors. We consider that the information assembled in this volume under the title, *Disease Suppressive Tropical Soils*, is the first in this area and is worth publishing, and it is our hope that many more researchers will show interest in this area.

The R&D Team member at Biotech Innovations, Andrea Borovic, receives special thanks and gratitude as they took heavy share of all the organizational work of the chapters. She is acknowledged for the exceptional help she rendered.

Contents

Chapter 1
Disease Suppressive Rice Soils in the United States and Their Importance to Crop Production

Samuel Gnanamanickam and Camila P. Nicolli

1 Introduction

The monoculture of rice is a well-known phenomenon which is associated with the history of rice production in the world. It is well known that monoculture of rice has an impact on disease pressures and such rice soils are known to have become disease suppressive. In this chapter, we describe the detailed studies that have been made in the Mekong delta of Vietnam and in Kerala of southern India (Chap. 5). The impact of this development of disease suppressive soils leads to environmentally friendly rice production and reduces the harmful effects of excessive application of pesticides and fungicides. More significantly, rice farmers spend less on the cost of cultivation and harvest 15–20% higher yields in Kerala (southern India) (Chap. 5). We record here how the rice farmers of Arkansas harvested 30–40% higher yields in their fields treated with microbial products during the years 2023 and 2024.

2 Detailed Studies

Rice Soils of Mekong Delta, Vietnam Hofte and her colleagues, Lam et al. (2021) of Ghent University, Belgium *published* how biological rice blast control takes place in highly acid sulfate soils of Mekong delta area of Vietnam. They reported that *Bacillus* cyclic lipopeptides Iturin and Fengycin controlled rice blast caused by

S. Gnanamanickam
R&D, Biotech Innovations, Wylie, TX, USA

C. P. Nicolli (✉)
Extension Plant Pathology, University of Arkansas System Division of Agriculture, Rice Research & Extension Center, Stuttgart, AR, USA
e-mail: sgnanamanickam@btilabs.com

S. Gnanamanickam, M. Höfte (eds.), *Disease Suppressive Tropical Soils*,
https://doi.org/10.1007/978-3-032-08100-1_1

Pyricularia oryzae in potting and acid sulfate soils by direct antagonism and induced systemic resistance. They observed that surfactin, iturin and fengycin completely suppressed *P. oryzae* spore germination resulting in disease severity reduction when co-applied on rice leaves (Lam et al. 2021).

Impacts of Monoculture and Disease Pressure on Rice Yield For two important reasons, this article by Lam et al. (2021) is significant: (1) The authors describe biological suppression of rice blast and (2) It is in a flooded delta area where soil is highly acidic. Lam et al. (2021) describe the acid sulfate soils (ASSs) of Mekong delta area in Vietnam: Acid sulfate soils account for more than 1.7% of the global cultivated land worldwide and are distributed over as much as 24 million hectares (Dent and Pons 1995; Faltmarsch et al. 2008). Most of these soils are cultivated with rice because of its tolerance to acidic conditions (Tanaka and Navasero 1966).

In Vietnam, more than 1.6 million hectares of ASSs in the Mekong Delta have been ameliorated for rice cultivation by managing submergence or drainage, applying phosphate fertilizer and lime, and using resistant cultivars (Olivier et al. 2000). Rice monoculture in ASSs has recently been intensified with three cropping seasons per year, instead of the single or double cropping seasons of the past decades. This leads to a higher disease pressure, resulting in a drastic reduction in rice yield. A prevailing disease which ravages rice fields and reduces annual rice production, especially in Vietnam, is the foliar rice blast disease caused by *Magnaporthe oryzae* (Lam et al. 2021).

2.1 The Kuttanad Rice Production Area in Kerala in Southern India

The above description of acid-sulfate soils (ASSs) is like what we are about to describe in the Kuttanad rice-growing region of Kerala in India (Muralidharan et al. 2017) (Chap. 5). In Kuttanad, the rice field soils were under complete flooding in July 2022 when Gnanamanickam visited this rice-growing area.

The farmers patiently wait for the water level to go down and be drained before starting the rice planting operation in November. Here is a photograph of the rice fields of Kuttanad (on 28th of July 2022) (Fig. 1.1):

2.2 Use of **Pseudomonas fluorescens** for Eco-Friendly Rice Farming

Gnanamanickam (1983) introduced *Pseudomonas fluorescens* in India for the biological control of plant diseases and later (1990–1992) specifically to control rice diseases, in particular rice blast disease (Gnanamanickam and Mew 1989, 1992). In

Fig. 1.1 Kuttanad, Kerala: A lake- like appearance. Rice fields are under standing water (as seen on July 28, 2022). This rice production area is under mean sea level. *(Photo by S. Gnanamanickam, July 2022)*

first reports, Velusamy and Gnanamanickam (2003) and Velusamy et al. (2006) observed the production of 2,4-diacetylphloroglucinol (DAPG) by the Indian strains of *P. fluorescens* collected from the rice paddies of southern India. There is, therefore, a high probability that rice soils in Kuttanad area of Kerala in southern India have become disease-suppressive due to progressive application of *P. fluorescens* since 2011.

In our opinion, what Muralidharan et al. (2017) wrote and recorded is the most outstanding example for the use of *Pseudomonas fluoresens* in a very flooded, and possibly highly acid sulfate soils. The articlce, "Resource conserving and eco friendly technologies for climate resilience and higher profitability of rice production in Kuttanad region of Kerala" (Muralidharan et al. 2017) describes how rice farmers of Kuttanad, Kerala have been using the technology for the last 13–14 years and Gnanamanickam who introduced *P. fluorescens* to India back in 1983 (Unnamalai and Gnanamanickam 1983) recently visited with leading rice farmers in this region in July 2022 and Dember, 2023 and had a personal interaction with farmers who apply *P. fluorescens* as seed treatment and follow up with two foliar sprays. **They do not apply pesticides or chemicals**.

Resource-conserving Technologies in Climate-Resilient Agriculture This article by Muralidharan et al. (2017) lists soil acidity as the No.1 major constraint for rice farming in the area. Rice farmers do apply dolomite (lime) along with *Pseudomonas* and trichocards (bio-insecticide) as advised by the ICAR (Indian Council of Agricultural Research) officials who have been co-ordinating this effort. This has become a highly note-worthy package of practices adopted by rice farmers of Kuttanad region and newspapers (English and Malayalam language) have written articles about this way of rice cultivation.

The Kuttanad rice growing area is a unique rice-farming region designated as **Globally Important Agricultural Heritage System (GIAHS) by FAO in 2013.**

The Paper: "Resource conserving and eco-friendly technologies for climate resilience and higher profitability of rice production in Kuttanad region of Kerala" (Muralidharan et al. 2017) is a detailed write-up and it is covered in much detail in Chap. 5.

3 Disease Suppressive Soils: A Well-Known Example of Take-All Decline (TAD) in Wheat

Role of *Pseudomonas* spp. and 2,4-diacetylphloroglucinol Take-all, caused by *Gaeumannomyces graminis* var. *tritici*, is the most economically important root disease of wheat worldwide. Take-all decline (TAD) is the spontaneous decline in incidence and severity of disease after a severe outbreak of take-all during continuous wheat or barley monoculture. TAD occurs worldwide, and in the United States and The Netherlands it results from a build-up of populations of 2,4-diacetylphloroglucinol (2,4-DAPG)-producing fluorescent *Pseudomonas* spp. during wheat monoculture. The antibiotic 2,4-DAPG has a broad spectrum of activity and is especially active against the take-all pathogen. Based on genotype analysis by repetitive sequence-based-PCR analysis and restriction fragment length polymorphism of *phlD*, a key 2,4-DAPG biosynthesis gene, has been found in at least 22 genotypes of 2,4-DAPG producing fluorescent *Pseudomonas* spp. These have a worldwide occurrence, including India. Velusamy and Gnanamanickam (2003) and Velusamy et al. (2006) observed the production of 2,4-diacetylphloroglucinol (DAPG) by the Indian strains of *P. fluorescens* collected from the rice paddies of southern India.

In a review, Kwak and Weller (2013) provided an overview of *G. graminis* var. *tritici*, the take-all disease, *Pseudomonas* biocontrol agents, and mechanism of disease suppression.

There is, therefore, a high probability that rice soils in Kuttanad area of Kerala in southern India have become disease-suppressive due to progressive application of *P. fluorescens* since 2011.

3.1 Monoculture of Wheat and Barley Makes Disease Suppressive Soils

History: Wheat has been cultivated in Washington since the 1820's. Studies in Washington State and the Netherlands have demonstrated that TAD develops because of the buildup of strains in the *Pseudomonas fluorescens* complex (Loper et al. 2012) that produce the broad-spectrum antibiotic 2,4-DAPG (PhID+) during monoculture of wheat or barley (Raaijmakers and Weller 1998, 2001; Weller et al. 2007).

Case Studies in Disease Suppressive Soils Like wheat monoculture in temperate regions of the world, rice monoculture is a well-known phenomenon in tropical areas of the world.

We choose to highlight the rice history in Kerala, the southern state of India. It is known that rice has been cultivated since time immemorial that is, a time that is beyond the reach of our memory (Please see the photograph showing the rice history) (Fig. 1.2).

Therefore, we the authors very strongly believe that in areas of the country where rice has been grown for hundreds and thousands of years, such as the Kuttanad rice-growing region of Kerala, application of *P.fluorescens* or PGPR strains by rice farmers leads to formation of soils that will progressively become suppressive to rice blast, sheath blight, bacterial leaf blight and other major rice diseases. From Gnanamanickam's laboratory, tropical strains of *P. fluorescens* that produced 2,4-diacetylphloroglucinol (DAPG)(the same antibiotic shown to have a major part in the Take-All Decline (TAD) of wheat in temperate areas of the world) were identified by Velusamy and Gnanamanickam (2003) and the antibiotic was shown as the major mechanism in the biological suppression of bacterial blight of rice (a major rice disease caused by *Xanthomonas oryzae* pv. *oryzae*) by Velusamy et al. (2006) in India.

3.2 Example of Successful Use of Microbial Products for Rice Production in the United States

Rice Production in the United States Each year, American rice farmers sustainably grow roughly 20 billion pounds of rice in Arkansas, California, Louisiana, Mississippi, Missouri, and Texas. Half of the rice crop is consumed in the United States, accounting for 80% of the rice consumed domestically.

Fig. 1.2 The Regional Agricultural Research Station in Pattambi, Kerala. The station has a long history and is established as a Paddy Breeding station by the British in 1927. Studies have shown that rice has been cultivated since time immemorial. Definition of immemorial: extending or existing since beyond the reach of memory. *(Photo by S. Gnanamanickam)*

The other half of the crop is exported to more than 120 countries around the globe. The U.S. rice industry provides critical jobs in rural areas throughout the country. As a whole, the rice industry contributes more than $34 billion to the U.S. economy annually and provides jobs for more than 125,000 individuals in the United States.

(Source: USA Rice Federation; https://www.usarice.com › thinkrice › discover-us-rice).

Trials with Microbial Products in Arkansas Gnanamanickam has managed the Research & Development research for a biotech company called Biotech Innovations (BTI) in Texas. BTI has been in the U.S. since 2010 and has a strong presence in the Asian Pacific market for over 30 years. Bti performs protein encapsulation of microbes for protection and longevity and the company, originally supplied independent distributors and sold under various labels in the U.S. BTI formulates, develops, manufactures microbial products and has a new expanded facility in Plano, TX.

Fig. 1.3 BTI biological products integrated into rice trial in Arkansas. *(Created by S. Gnanamanickam)*

Farmer Observations and Yield Improvements During the year 2023 and 2024, yield trials were scheduled in Arkansas in a selected number of rice farms, with the following objectives: to measure the efficacy of BTI biological products in Arkansas rice production (Nicolli, C. 2024: Personal observations of greenhouse experiments).

- 5 field-level trials were conducted across northeast Arkansas
- All the fields were in the flooded rice production systems.
- Anheuser-Busch corporation's field agronomy team coordinated these field trials (Fig. 1.3).

There were noticeable differences in the health and vigor of the treated rice as shown below (Fig. 1.4):

Therefore, this response of rice to microbial products or biological products created a very positive attitude in the rice farmers to the use of microbial products in rice production.

During the year 2024, again the same 5 rice farms were selected, and the yield trials were repeated using the same protocols. When rice was in the farms, Gnanamanickam and colleagues visited the rice farms in Jonesboro, Arkansas. The rice farmers were even happier this time with their crops. One of those farmers, Zack Macmillan's crop is shown below (Fig. 1.5):

3.3 Conclusion

From data analysis of the Arkansas rice yield trials, the following conclusions were made (Fig. 1.6):

Fig. 1.4 Compton rice farm: the treated rice yielded 214 bushels of rice/ac while the untreated rice yielded 184 bushels of rice/ac. Thus, the increase in grain yield is 30 bushels/ac. *(Photo by S. Gnanamanickam)*

Fig. 1.5 Harvest of 30–40% higher grain yield during 2024 in Arkansas. *(Photo by S. Gnanamanickam)*

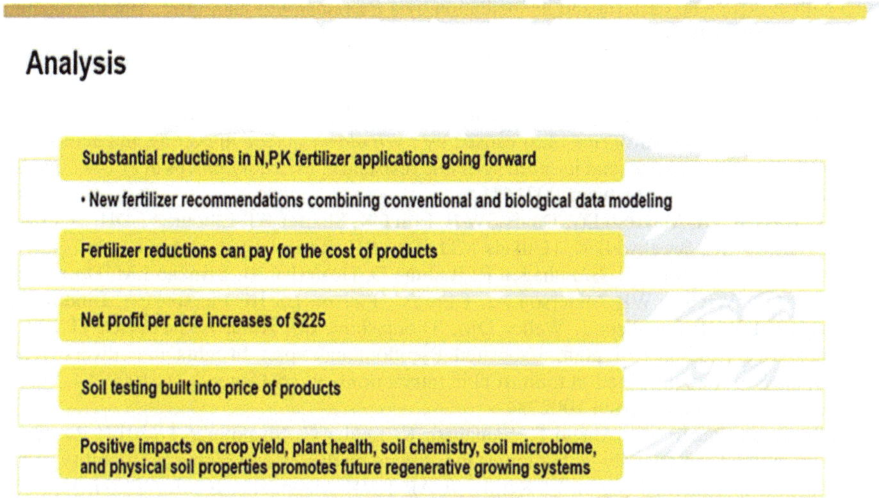

Fig. 1.6 Data analysis. *(Made by S. Gnanamanickam)*

3.4 Disease Suppressive Soils: How They Affect Crop Production in Other Crops

In this present volume, there are chapters or contributions from well-known authors/institutions of China (Chap. 4), India (Chap. 5), Malaysia (Chap. 6) and other areas of the Tropics whose research on other crops (Cocoyam, Chap. 2)) reveal how disease suppressive soils is the important key for crop production of banana, or rice or oil palm. These contributions tell us further about the significant interactions between rhizobacteria and crop plants that confer salinity and drought stress tolerance in the Tropics.

The concept that has emerged is the Disease Suppressive Soils and how they affect crop production in the Tropics. This was featured in the new volume on Rice Blast Disease (Gnanamanickam 2023). Also, this is what led to the development of the present volume on Disease Suppressive Tropical Soils.

References

Dent DA, Pons L (1995) World perspective on acid sulphate soils. Geoderma 67:263–276

Fältmarsch RM, Åström ME, Vuori KM (2008) Environmental risks of metals mobilized from acid sulphate soils in Finland: a literature review. Boreal Environ Res 13:444–456

Gnanamanickam SS (2023) Rice Blast Disease: contributions to biological disease management and the Biopesticide Industry. Copyright: Samuel S Gnanamanickam. 224p

Gnanamanickam SS, Mew TW (1989) Biological control of rice blast with antagonistic bacteria. Int Rice Res Newslett 14:34–35

Gnanamanickam SS, Mew TW (1992) Biological control of blast disease of rice (*Oryza sativa*, L) with antagonistic bacteria and its mediation by a *Pseudomonas* antibiotic. Ann Phytopathol Soc Japan 58:380–385

Kwak Y-S, Weller D (2013) Take-All of wheat and natural disease suppression: a review. Plant Pathol J 29:125–135

Lam VB, Meyer T, Aria AA, Ongena M, Oni FE, Hofte M (2021) *Bacillus* cyclic lipopeptides iturin and fengycin control rice blast caused by *Pyricularia oryzae* in potting and acid sulfate soils by direct antagonism and induced systemic resistance. Microorganisms 9(7):1441. https://doi.org/10.3390/microorganisms9071441

Loper JE, Hassan KA, Mavrodi DV, Davis EW II, Lim CK, Shaffer BT, Elbourne LDH, Stockwell VO, Hartney SL, Breakwell K, Henkels MD, Tetu SG, Rangel LI, Kidarsa TA, Wilson NL, van de Mortel JE, Song C, Blumhagen R, Radune D, Hostetler JB, Brinkac LM, Durkin AC, Klupfel DA, Wechter WP, Anderson AJ, Kim YC, Pierson LS III, Pierson EA, Lindow SE, Kobayash DY, Raaijmakers J, Weller DM, Thomashow LS, Allen AE, Paulsen IT (2012) Comparative genomics of plant-associated Pseudomonas spp.: Insights into diversity and inheritance of traits involved in multitrophic interactions. PLoS Genet 8(7):e1002784. https://doi.org/10.1371/journal.pgen.1002784

Muralidharan P, Rajeev MS, Anand R, Sajnanath K, Lekha G, Sivakumar T (2017) Resource-conserving and eco-friendly technologies for climate resilience and higher profitability of rice production in Kuttanad region of Kerala. ICAR-KVK-Alappuzha, ICAR-CPCRI, Regional Station, Kayankulam, p 12p

Olivier H, Hanhart K, Phung M, BoumaWater J (2000) Management for rice cultivation on acid sulphate soils in the plain of reeds, Vietnam. Agric Water Manag 46:91–109

Raaijmakers J, Weller DM (1998) Natural plant protection by 2,4-diacetylphloroglucinol-producing *Pseudomonas* spp. in take-all decline soils. Mol Plant-Microbe Interact 11:144–152

Raaijmakers J, Weller DM (2001) Exploiting diversity of 2,4-diacetylphloroglucinol-producing *Pseudomonas* spp.: characterization of superior root-colonizing *P. fluorescens* strain Q8r1-96. Appl Environ Microbiol 67:2545–2554

Tanaka A, Navasero SA (1966) Growth of the rice plant on acid sulfate soils. Soil Sci Plant Nutr 12:23–30

Unnamalai N, Gnanamanickam SS (1983) *Pseudomonas fluorescens* is an antagonist to *Xanthomonas citri* (Hasse) Dye, the incitant of citrus canker. Curr Sci 50:703–704

Velusamy P, Gnanamanickam SS (2003) Identification of 2,4-diacetylphloroglucinol (DAPG) production by plant-associated bacteria and its role in suppression of rice bacterial blight in India. Curr Sci 85:1270–1273

Velusamy P, Emmanuel JE, Gnanamanickam SS, Thomashow LS (2006) Biological control of rice bacterial blight by plant-associated bacteria producing 2,4- diacetylphloroglucinol. Can J Microbiol 52:56–65

Weller DM, Landa BB, Mavrodi OV, Schroeder KL, De La Fuente L, Blouin Bankhead S, Allende Molar R, Bonsall RF, Mavrodi DV, Thomashow LS (2007) Role of 2,4-diacetylphloroglucinol-producing fluorescent *Pseudomonas* spp. in the defense of plant roots. *Plant Biol (Stuttg)* 9:4–20

Chapter 2
Cocoyam Root Rot Disease Caused by *Pythium myriotylum*: Etiology, Epidemiology and Management

Feyisara Eyiwumi Oni, Amayana Adiobo, Joseph Onyeka, and Monica Höfte

1 Introduction

Cocoyam (*Xanthosoma sagittifolium* (L.) Schott), belonging to the family Araceae, is one of the six most important root and tuber crops grown worldwide (Onwueme and Charles 1994). Next to cassava and yam, cocoyam ranks third in the tropics as staple energy food. The corms, cormels and leaves are a vital source of carbohydrates, vitamins and minerals for human nutrition and animal feed (Boakye et al. 2018; Nyochembeng and Garton 1998). Although native to South America, it has spread to the West Indies, Southeast Asia, the Pacific Islands and tropical Africa (Acevedo-Rodriguez and Strong 2012; Govaerts et al. 2002; PIER 2022; PROTA4U 2025). In spite of the fact that roughly 400 million people worldwide rely on cocoyam for survival (Onokpise et al. 1999), it is grossly under-researched mainly because the cocoyam crop is consumed where it is produced and is not under the mandate of the international research system (Lebot et al. 2024).

F. E. Oni
Laboratory of Phytopathology, Department of Plants and Crops, Ghent University, Coupure Links, Ghent, Belgium

Phytopathology Department, Rijk Zwaan Breeding B.V, De Lier, The Netherlands

A. Adiobo
Jay P J Biotechnology Laboratory, Institute of Agricultural Research for Development (IRAD), Ekona, Buea, Cameroon

J. Onyeka
Plant Pathology Unit, National Root Crops Research Institute, PMB, Umuahia, Nigeria

M. Höfte (✉)
Faculty of Bioscience Engineering, Ghent University, Gent, Belgium
e-mail: monica.hofte@ugent.be

S. Gnanamanickam, M. Höfte (eds.), *Disease Suppressive Tropical Soils*,
https://doi.org/10.1007/978-3-032-08100-1_2

Fig. 2.1 (**a**) *Xanthosoma sagittifolium*, white variety; (**b**) *Xanthosoma sagittifolium*, red variety; (**c**) *Colocasia esculenta* (with symptoms of taro blight caused by *Phytophthora colocasiae*)

The Araceae is a huge family of herbaceous monocots most of which occur in the New World tropics (Boyce and Croat 2011 onwards). Of its numerous members, *X. sagittifolium* (otherwise known as tannia) and *Colocasia esculenta* (popularly known as taro or dasheen) are attaining world importance as energy foods (Giacometti and León 1994). At first glance, a *Colocasia* plant could be mistaken for *Xanthosoma*, however, a closer look enables spotting out differences in the leaf morphology (Fig. 2.1). Although in both genera, the leaf tips point downward, the petiole attachment of *Xanthosoma* is at the leaf notch whereas that of *Colocasia* is below the notched edge of the leaf. Additionally, the leaf lamina of *Xanthosoma* is often sagittate (arrow-shaped) while that of *Colocasia* is cordate (heart-shaped). More so, a prominent vein which runs along the margin of the *Xanthosoma* leaf is conspicuously absent in taro (Onwueme and Charles 1994). In this review, our focus is on a member of the *Xanthosoma* genus, the cocoyam, taxonomically named *Xanthosoma sagittifolium*.

The *Xanthosoma* genus is divided in species based on colour of corm and cormels, cormel shape and the leaves (Brown 2000; Quero-Garcia and Lebot 2010) but the taxonomy is confusing and in need of revision (Lebot et al. 2024). In the KEW database 198 accepted species names are listed (POWO 2025) but many could be synonyms. Members of this genus are grown either as ornamentals or food crops (Quero-Garcia and Lebot 2010). Among these, *Xanthosoma sagittifolium* (L.) Schott is the most widely grown (Onwueme 1978). There is a tendency to give the name of *Xanthosoma sagittifolium* to all cultivated edible *Xanthosoma* species (Giacometti and León 1994), although a number of other edible species in this genus have been recognized (Lebot et al. 2024). However, some such as *X. violaceum, X. atrovirens, X. mafaffa*, and *X. roseum* are nowadays considered as synonyms of *X. sagittifolium* (POWO 2025). *X. brasiliense* is a small species cultivated solely for its edible leaves (Ramesh et al. 2007; Onwueme and Charles 1994). *X. caracu*, also known as yautia horqueta, is cultivated in Puerto Rico and both cornels and leaves are eaten. *X. robustrum* is grown in Mexico and Costa Rica. The native range of *X. taioba* is Brazil where it is eaten as a leafy vegetable (Gomes de Souza et al. 2023).

It is suggested that cocoyam is native to South America including Colombia, Ecuador, Peru and Venezuela where it has been cultivated since prehistoric times by the Mayan Indians (Govaerts et al. 2002; Onwueme and Charles 1994) long before

Europeans arrived in the fifteenth century. From there, it spread to the West Indies, Southeast Asia, the Pacific Islands and tropical Africa (Acevedo-Rodriguez and Strong 2012; Govaerts et al. 2002; PIER 2022; PROTA4U 2025). In the sixteenth and seventeenth century, *Xanthosoma* reached West Africa when it was loaded on board ships heading for the slave coast to feed the human cargo (Brown 2000). From the ports *Xanthosoma* was further spread by traders and missionaries and it was locally named new cocoyam since it replaced the 'old' cocoyam (*Colocasia esculenta* (L.) Schott), an aroid root crop from Southeast Asia. Eventually, cocoyam was introduced into the South Pacific and Asia by Europeans in the nineteenth century. Cocoyam fields are a common sight in Florida, in response to the needs of the large population of immigrants from South America and Africa (Brown 2000).

The cocoyam (*X. sagittifolium*) crop provides carbohydrates in human diet and also contains substantial amounts of lipids, proteins, and essential vitamins like beta-carotene, riboflavin, niacin, ascorbic acid, and thiamine (Boakye et al. 2018; Cobley and Steele 1976; Onwueme and Charles 1994). In West and Central Africa, which constitute the major cocoyam-producing regions, corms and cormels are boiled, roasted or baked but the most popular preparation method is by drying and grounding the cormels into a flour called 'fufu'. The leaves can be prepared as spinach in soups or salads (Adiobo et al. 2007; Boakye et al. 2018). The corm, cormels and sometimes, the leaves are the main economically important parts of the plant although the cormels are tasty and preferable to the corms. Since they are grown in diverse countries and cultures, cocoyam has a lot of given names including 'tannia' (Central and South America), and malanga (Spain) (Reyes Castro 2006). In West Africa, besides being named 'macabo', it is also called the 'new cocoyam' since it replaced the 'old' cocoyam (*Colocasia esculenta* (L.) Schott) (Perneel 2006).

There are at least three cultivated *Xanthosoma* varieties based on the colour of the flesh (Perneel 2006; Reyes Castro 2006). The white cocoyam ($2n = 26$), which is the most preferred variety for its high yielding and good taste, the yellow cocoyam ($2n = 52$) which does not tuberculize, and the red cocoyam ($2n = 26$) (Fig. 2.1). Cocoyam is an annual plant with growth cycle of 9–12 months and is vegetatively propagated by farmers via suckers since natural flowering is rare. In this review, we will refer to the cocoyam crop as the 'tannia'.

In spite of its potential to alleviate hunger and poverty for people living in the tropics, tannia production is limited by two devastating diseases. Dasheen mosaic virus (DMV) is the most important virus that infects cocoyam (Reyes et al. 2009; Reyes et al. 2006). Under field conditions, virus-free plants performed better with respect to yield in comparison with infected *in vitro* plants. Yield losses of up to 90% have been reported to occur due to the DMV (Reddy 2015). A second and important limitation is the destruction due to the Cocoyam Root Rot Disease (CRRD) caused by the oomycete, *Pythium myriotylum*, leading to losses as high as 90% in infected fields (Pacumbaba et al. 1992b). Interestingly, the root rot disease caused by *P. myriotylum* is specific to the *Xanthosoma* and does not infect the closely related taro tuber (*C. esculenta*) (Misra et al. 2008). In contrast, *C. esculenta* is highly susceptible to *Phytophthora colocasiae*, causing taro leaf blight (Brooks 2008; Onyeka 2014), while *X. sagittifolium* is not infected.

The objectives of this review are (i) to provide an overview of CRRD caused by *P. myriotylum* on *X. sagittifolium,* (ii) to discuss the identification, etiology, and epidemiology of the pathogen and disease, and (iii) to discuss management strategies with an emphasis on suppressive soils and biological control.

2 The Cocoyam Root Rot Disease

2.1 *Discovery and Background*

CRRD was reported for the first time in Ghana (Purseglove 1972). Subsequently, root rot of tannia was described as an economically important disease which limits cocoyam production in the lowland forest area of Cameroon (Steiner 1981). This disease was attributed to *P. myriotylum* Drechsl., which could be isolated from infected roots and provoked dwarfing of plants after artificial inoculation. Steiner fulfilled Koch's postulates after a successful re-isolation of the pathogen from roots. Similar results were shown by Nzietchueng (1983) and Nzietchueng (1984). In spite of these afore-mentioned independent studies, reports persisted that CRRD was caused by a pathogen complex, involving *Rhizoctonia solani* and *Fusarium solani* that are often associated with *P. myriotylum* in diseased cocoyam roots (Ofoegbu 1993; Théberge 1985). To clarify this, Pacumbaba et al. (1992b) isolated *P. myriotylum, F. solani* and *R. solani* from CRRD-infected plants in Cameroon. Pathogenicity test of these individual pathogens on 3- and 7-month old cocoyam plantlets showed that the root rot disease of cocoyams can be attributed solely to *P. myriotylum* and not to *F. solani* and *R. solani* or a combination of both. Thus, *F. solani* and *R. solani* were designated as opportunistic pathogens. Further studies to investigate CRRD showed comparable pathogenicity of *P. myriotylum* isolates obtained from Cameroon (Adiobo et al. 2007; Olorunleke et al. 2014; Perneel 2006; Perneel et al. 2006).

2.2 *Further Incidence and Distribution of the Disease*

In Costa Rica, the CRRD was initially reported as being caused by *Fusarium* sp. (Saborio et al. 2004). However, a few years later *P. myriotylum* was isolated from diseased cocoyam roots in Costa Rica. Furthermore, pathogenicity tests with selected isolates confirmed this pathogen is also the causal agent of CRRD in Costa Rica (Perneel et al. 2006). CRRD caused by *P. myriotylum* has also been reported in Sri-Lanka (Tojo et al. 2005) and Nigeria (Olorunleke et al. 2014).

2.3 Pathogen Identification and Biology

An efficient protocol for the isolation of *P. myriotylum* from cocoyam roots has been previously described (Perneel et al. 2006; Tambong et al. 1999). A scheme of this protocol is shown in Fig. 2.2. To isolate the pathogen, infected cocoyam roots with typical necrotic spots are collected from the soil. Roots are cut into sections of 3–4 cm, surface sterilized with 1% commercial bleach (0.053% NaOCl) for 2–3 min and blotted on filter paper. Trimmed sections are incubated at 28 °C on 1.5% water agar supplemented with 150–300 mg $l^{-1}$of streptomycin sulfate for 48 h. Actively growing hyphal tips are transferred aseptically to antibiotic-free water agar and incubated as previously done.

Microscopic characterisation of *P. myriotylum* shows that hyphae are straight, sinusoidal and not much branched except at the end. The pathogen can be identified by the presence of smooth sporangia which occur terminally or intercalary, consisting of long, simple portions of unswollen hyphae. Knob-like appressoria are often present appearing in branching clusters. Oogonia are abundant and mostly terminal, whereas three to six antheridia appear per oogonium with slender stalks and sometimes basal contact (Waterhouse and Waterston 1966). Zoospore formation can be induced from sporangia of detached mycelia grown on lima bean sucrose agar and soaked in 0.001 M sucrose. In addition to this, zoospores are reportedly produced in water containing high amounts of Ca^{2+} and an increased pH of up to 10.5 (Nyochembeng et al. 2002). Motile zoospores of *P. myriotylum* could also be induced by flooding mycelial mats of the pathogen with saccharose buffer, in Petri dishes (Pacumbaba et al. 1992b). The mycelial mats were spread by rotating the dish by hand, after which the plates were incubated at 31 °C in continuous

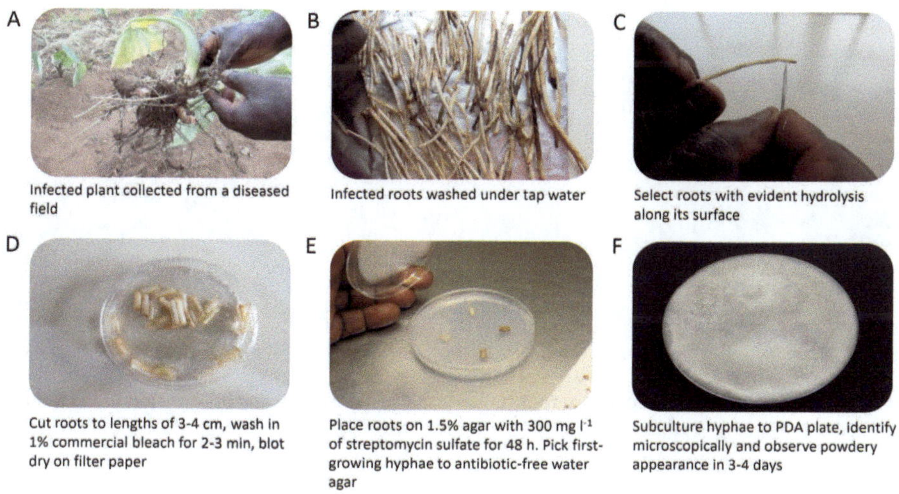

Fig. 2.2 Steps for the isolation of *P. myriotylum* from diseased cocoyam roots

fluorescent light for 20 h. Zoospores were harvested and the process could be repeated several times.

P. myriotylum isolates from cocoyam grow optimally at 28 °C, with no growth at 37 °C (Perneel et al. 2006). Young cultures of *P. myriotylum* have a typical powdery appearance on PDA medium and the mycelium becomes fluffy after 4 days. In contrast, *P. myriotylum* isolates from other hosts immediately produce fluffy mycelium but do not give the powdery appearance.

2.4 Signs and Symptoms

Symptoms reported included dwarfing of plants such that they produced small leaves which decay early (Fig. 2.3). Production of tubers is either suppressed completely or small tubers are derived. A second symptom is yellowing of leaves, which for example occurs in Cameroon at the onset of heavy rain. The root system is reported to decay due to rot within 3–4 weeks which further spreads in foci that could expand up to 5 m per week (Pacumbaba et al. 1992b; Perneel 2006; Steiner 1981).

2.5 Yield Losses

Yield reductions of up to 90% have been reported in some plantations in Cameroon (Nzietchueng 1983; Schafer 1999). The documentation of yield losses due to CRRD is not readily available for other countries.

Fig. 2.3 Symptoms of the cocoyam root rot disease caused by *Pythium myriotylum* in the field

3 Etiology of the Cocoyam Root Rot Disease

3.1 Ecology, Infection and Survival

The main sources of inoculum are infected soil and infected planting material. Furthermore, the pathogen is soil borne and disease development is aided by poorly drained soils. Additionally, *P. myriotylum* persists in the soil for many years (Nzietchueng 1984).

3.2 Disease Cycle

The primary inoculum for the propagation of CRRD originates from infected soil and infected propagation material (Adiobo 2006). In Cameroon for example, most cocoyam growers established their farms using planting material (suckers, sprouting fragments of corms) obtained from abandoned old fields. Cocoyam is intercropped with annual crops including beans, groundnuts and cowpeas, which are potential hosts of *P. myriotylum*, the causal organism of CRRD (Tambong et al. 1999). In spite of the evidence of host specialization of the CRRD pathogen, such crops may still act as alternative hosts (Adiobo 2006). Furthermore, cocoyam residues such as corms and unharvested tubers with a potential of producing volunteer plants are sometimes not properly managed in fields. Such slow-decaying organic materials could serve as a source of initial or secondary inoculum. Furthermore, improperly carried out sanitation practices of removing, rouging, and destroying of corms provide inoculum reservoirs thereby enhancing the spread of the disease (Adiobo 2006).

Initial infection by *P. myriotylum* occurs when the pathogen directly penetrates the host by an appressoria (Drechsler 1930). Pathogen attack begins with the colonization of the peripheral and epidermal cells of the root apex (Boudjeko et al. 2006). These cells become rapidly lost after infection while the cortical and stele cells remain. Further studies suggest that during the early stage of infection, the oomycete causes a significant loss of pectin probably through degradation by hydrolytic enzymes which diffuse and act away from the site of attack. Loss of pectin was evidenced by a significant decrease in galacturonic acid content within the cell walls of infected roots. On the other hand, the crop responds in the early stages of infection by the formation of wall appositions which are rich in callose and cellulose (Boudjeko et al. 2006).

If infected at the emerging roots stage, about 1–2 months following planting, stunting of cocoyam plants occurs (Adiobo 2006). More so, *P. myriotylum* can still infect well established cocoyam plants aged 3–5 months. The infected plant develops chlorotic leaves followed by brown blight that starts at the periphery of the leaf and extends towards the petiole. Thereafter, leaves become yellow, dry up and then shed prematurely. The root rots, the cortical tissue turns brown and disintegrates leaving a non-functional vascular skeleton. In severe attacks, the entire root is destroyed (Adiobo 2006).

In the asexual cycle, the mycelium produces sporangia, which germinate to produce vesicles containing zoospores. These zoospores encyst and germinate to produce a germ tube capable of host penetration or production of another sporangia. Sexual reproduction occurs via mating of an oogonia and antheridia which after fertilization produces oospores that are capable of overwintering thereby perpetuating the sexual cycle (van West et al. 2003). Zoospores are released during wet conditions, transported in free water and able to swim to new infection sites; therefore, wet conditions are required for disease to occur (van West et al. 2003). Invasion by *P. myriotylum* propagules in the detached roots of cocoyam either by zoospores or mycelial strands was observed to be intra- and intercellular. The process occurred 6–7 h following contact of the pathogen with the detached roots at 31 °C. After 7 h, the rotting of the roots commence (Pacumbaba et al. 1994).

3.3 Detection

Pythium myriotylum is a soil-borne plant pathogen with a cosmopolitan distribution and able to incite severe root damage on a wide variety of crops (broad host range). *P. myriotylum* isolates from cocoyam can be differentiated from other isolates by their optimum growth temperature, esterase patterns, AFLP fingerprints, and sequences of the internal transcribed spacer regions (ITS2) located between the 5.8S and 28S ribosomal DNA (rDNA-ITS2) (Perneel et al. 2006; Tambong et al. 1999). Furthermore, Gómez-Alpízar et al. (2011) successful developed a PCR-RFLP method using specific primers to distinguish cocoyam-pathogenic *P. myriotylum* isolates from non-pathogenic ones.

3.4 Pathogen Diversity and Host Adaptation

Pythium species seldom exhibit host specificity and isolates within species vary in pathogenicity and virulence on varieties of specific crops (Moorman et al. 2002). However, *P. myriotylum* on cocoyam fields in Cameroon appeared to show more severe symptoms on cocoyam than what was observed in other host plants such as groundnuts and beans (Tambong et al. 1999). Following this, Perneel et al. (2006) investigated the possibility for intraspecific variability between *P. myriotylum* isolated from cocoyam and those obtained from other host crops. Results of this study gave biochemical and molecular evidence that cocoyam-pathogenic *P. myriotylum* isolates have developed a certain degree of host specialization (Perneel et al. 2006). Analysis of esterase banding patterns showed that unlike *P. myriotylum* isolates from other host crops, isolates from cocoyam did not produce any monomorphic markers. Furthermore, alignments of ITS sequences showed a base transition at position 824 from adenine in typical isolates of *P. myriotylum*, to guanine, in isolates that were pathogenic on cocoyam. Additionally, only isolates with guanine in position 824 of the ITS sequences could infect cocoyam. Together with AFLP analysis, these results highlight the host adaptation of cocoyam-pathogenic *P. myriotylum* (Perneel et al. 2006). Furthermore, multiple sequence analysis was conducted using *P. myriotylum* isolates obtained from cocoyam and other host crops including sorghum, pineapple, beans and snuff. Based on sequencing and analysis of cytochrome oxidase I, cytochrome oxidase II and the β-tubulin gene sequences, additional evidence was obtained for the speciation of *P. myriotylum* in cocoyam. Subsequently, the name *P. myriotylum* var. *aracearum* was proposed in order to distinguish these cocoyam-pathogenic isolates from other *P. myriotylum* isolates (Tambong et al. 2006).

4 Epidemiology of the Cocoyam Root Rot Disease

In general, soil moisture, water logged soil conditions, high soil temperature and high planting density favour development of CRRD (Nzietchueng 1985). High planting density provided favorable conditions for a fast spread of root rot of cocoyam.

Studies on the epidemiology of the CRRD were conducted in Cameroon for three consecutive years; 1979–1981. Experiments were conducted in the rainy season, from the middle of March to the end of November and in volcanic soil. CRRD was spread through infested soil or infested planting material and could persist in the soil for many years (Nzietchueng 1985). Plants from infested corms planted in infested soil developed poorly and remained dwarfed throughout their vegetative cycle whereas those from either infested soil with healthy corms or from infested corms with pathogen-free soil developed in most of the cases a yellowing symptom on the leaves.

Disease was observed to spread in foci that developed at the same time on the field. Most of the foci started from water-logged sites in the field. Mapping of the disease showed that the rate of the spread was relatively fast. Foci could expand 6–8 m per week and secondary foci developed afterwards. Because the spread of the disease is from many foci, it was not always easy to establish the rate of the spread. Observations on disease evolution showed that rapid multiplication of the pathogen in June corresponded with optimum environmental parameters such as mean temperature (26 °C), relative humidity (89%) and rainfall (1800–2500 mm in July–September). Furthermore, experiments conducted in Cameroon showed that low population planting densities slow CRRD development and vice-versa.

5 Management of the Cocoyam Root Rot Disease

Fungicides have not been fully effective against CRRD (Nzietchueng 1983). Wide spacing, high mounds, regulation of time of planting (Onwueme and Charles 1994), drainage improvement, use of disease-free planting material, planting on ridges, crop rotation (Giacometti and León 1994), and the use of organic fertilizers have all been suggested in order to control the disease (Reddy 2015).

5.1 Appropriate Cultural Practices (Tillage, Fallow)

To salvage their crops from CRRD, cocoyam farmers engage in cultural practices including multiple cropping, hill planting, early planting, and early harvesting (Onokpise et al. 1999). Employment of wide crop spacing, regulation of the time of planting (Onwueme and Charles 1994), use of disease-free planting material using meristem tissue culture (Onokpise et al. 1992), crop rotation (Giacometti and León 1994), the use of organic manures (Reddy 2015), long interval of bush fallowing and roughing of diseased plants (Adiobo et al. 2007). A field survey conducted in Cameroon which involved data collection and analysis based on farmers' interviewed indicated the effect of several factors on CRRD including site selection, plot cropping history, land preparation, origin and quality of planting material, planting time, cropping system, soil management, and sanitation practices (Adiobo et al. 2007).

5.2 Chemicals

Although some water soluble fungicides such as metalaxyl suppress *Pythium* sp., they cannot provide effective control for diseases caused by this genus unless if applied before the plant is infected. Records of successful chemical control in green houses have been attributed to the limited amount of soil which allows for complete soil disinfection before planting (Adiobo 2006). Besides, field control of *Pythium* sp. has been considered to be difficult since fungal propagules are not exposed to the chemical coupled with the fact that the effect of most chemicals are short lasting such that the pathogen can easily survive via production of oospores or chlamydospores which are less sensitive to fungicides.

Field studies in Cameroon showed that high doses of metalaxyl (10 g/cocoyam plant) could reduce CRRD incidence by 80% and that higher doses (20 g/plant) resulted in phytotoxicity (Nzietchueng 1985). Additionally, the use of chemicals against CRRD were shown to be less successful (Onokpise et al. 1999; Pacumbaba et al. 1992a). The fungicide, Ridomil Plus 72, used to be effective in the control of the pathogen but was not cost-effective nor readily available to farmers (Onokpise et al. 1999). In the light of these limitations, alternative control strategies are in great demand.

5.3 Biological Control

Biological control against the CRRD has mainly been demonstrated using *Pseudomonas* and *Trichoderma* isolates. Fluorescent pseudomonads have long been recognized as effective biocontrol agents against plant pathogens (Höfte 2021; Höfte and Altier 2010; Olorunleke et al. 2015). The most commonly reported mechanism of biocontrol by fluorescent pseudomonads are via the production of cyclic lipopeptides, antibiotics, hydrogen cyanide and lytic enzymes, competition for nutrient and niches, competition for iron-mediated siderophores and induced systemic resistance. More so, many fluorescent pseudomonads have the capacity to stimulate plant growth by enhancing the availability and uptake of mineral nutrients through employment of phosphate-solubilization enzymes (Höfte and Altier 2010).

Pseudomonas aeruginosa PNA1, isolated from chickpea roots in India, showed strong biocontrol efficacy against Fusarium wilt on chickpeas and pigeonpeas (Anjaiah et al. 1998) and also against *Pythium splendens* damping-off of common bean (Anjaiah et al. 2003). Using *in vitro* and sterile volcanic soil experiments, Tambong and Höfte (2001) compared the biocontrol efficacy of *P. aeruginosa* PNA1 and its phenazine mutant against CRRD. Dual plate assays coupled with growing *P. myriotylum* in potato dextrose broth amended with bacterial supernatants, showed that *P. aeruginosa* PNA1 significantly reduced mycelial growth in the former whereas, mycelial dry weight was significantly reduced in the latter. Results obtained from plant experiments further highlighted the disease suppressiveness of PNA1 and the involvement of phenazines in the biocontrol capacity of this strain. Additively, Perneel et al. (2008) demonstrated the synergistic role of phenazines and rhamnolipids biosurfactants, produced by *P. aeruginosa* PNA1, against *P. myriotylum* on cocoyam. Given that *P. aeruginosa* is a clinical pathogen, it was imperative to search for alternative *Pseudomonas* biocontrol strains which produce both phenazines and biosurfactant that could possess comparable biocontrol efficacy as *P. aeruginosa* PNA1.

In Cameroon, fluorescent *Pseudomonas* strains were randomly isolated from the rhizosphere of healthy red and white cocoyam in naturally infested CRRD fields (Perneel et al. 2007) All *Pseudomonas* isolates were screened for *in vitro* antagonism against *P. myriotylum* in dual cultures. Besides the observation of field tolerance by red cocoyam towards CRRD, this study revealed several interesting results. Among the *Pseudomonas* isolates tested, several CRRD antagonists were found, all of which were isolated exclusively from the red cocoyam rhizosphere. Interestingly, all except one produced phenazine antibiotics. Furthermore, the efficacy of two of these strains, *Pseudomonas* sp. CMR12a and CMR5c, against CRRD was compared with that of *P. aeruginosa* PNA1 during plant experiments. While *P. aeruginosa* reduced CRRD severity by 48%, Pseudomonas sp. CMR12a and CMR5c could reduce root rot severity by 60% and 53%, respectively (Perneel et al. 2007). In a similar study, the well-studied phenazine producing strain *P. chlororaphis* PCL1391 (Chin-A-Woeng et al. 1998) also showed excellent control of the CRRD (Perneel 2006).

In follow-up studies, CMR12a (D'aes et al. 2014) and CMR5c (Ma et al. 2016) were shown to produce secondary metabolites including phenazine antibiotics and cyclic lipopeptides (CLPs). Specifically, *Pseudomonas* sp. CMR12a for example, produces two cyclic lipopeptides (orfamides and sessilins) in addition to phenazines (D'aes et al. 2011; D'aes et al. 2014). In this strain, mutants deficient in at least one, two or all three secondary metabolites were created (D'aes et al. 2014). Subsequently, biocontrol tests employing the wild type and mutant strains of CMR12a highlighted the roles of all three metabolites in the biocontrol capacity of this strain against *Pythium*-mediated CRRD (Oni et al. 2019a). Both strains belong to new species within the *P. protegens* subgroup. They have been described as *P. sessilinigenes* CMR12a and *P. aestus* CMR5c (Girard et al. 2021). The name *P. aestus*, however, is not validly described and is synonymous to *P. piscis*, so the correct name is *P. piscis* CMR5c (Mulet et al. 2023). In another follow-up study, fluorescent *Pseudomonas* strains obtained from healthy red and white cocoyams from three different fields in Cameroon (Boteva, Ekona and Maumu) were taxonomically characterized and their potential to produce CLPs was assessed to test the hypothesis that the more disease tolerant red cocoyam genotype selects for specific *Pseudomonas* strains which could contribute to the observed field tolerance to CRRD. There was no clear effect of cocoyam genotype on *Pseudomonas* spp. diversity, but the red cocoyam rhizosphere appears to support more diverse CLP producers than the white cocoyam rhizosphere. Eleven different CLPs were found on the red cocoyam whereas seven different CLPs were produced by *Pseudomonas* strains obtained from the white cocoyam. All CLPs showed effective antagonism against *P. myriotylum* during *in vitro* tests and a set of representative isolates strongly suppressed CRRD and effectively colonized cocoyam roots at high densities in plant experiments (Oni et al. 2019b).

Further biocontrol assays with *P. myriotylum* utilized four *Trichoderma* strains namely, *T. asperellum* PR10 and PR11, PR12 and 659–7 (Mbarga et al. 2012). They were previously isolated from cocoyam tubers in Cameroon and showed mycoparasitism against *Phytophthora megakarya*, the causal agent of black pod disease of cacao (Tondje et al. 2007). Initial dual culture *in vitro* tests using the four *T. asperellum* isolates revealed mycoparasitism and potential for biocontrol. Subsequently, *T. asperellum* PR10 and PR11 were used in plant experiments and both isolates could reduce *P. myriotylum* infection on the white cocoyam variety by 50% (Mbarga et al. 2012). Four additional *Trichoderma* isolates, obtained from a culture collection maintained at the University of Douala in Cameroon were tested for *in vitro* activity *against P. myriotylum*. The four strains, characterized as *T. erinaceum* IT 58, *T. gamsii* IT-62, *T. afroharzianum* P8 and *T. harzianum* P11 all inhibited mycelium growth of *P. myriotylum in vitro* which could be due to the combined effect of lytic enzymes, volatile and non-volatile organic compounds (Tchameni et al. 2020).

5.4 Disease Suppressive Soils

In some soils, indigenous plant-beneficial microorganisms effectively protect plants from certain plant pathogens, and these soils are referred to as being disease suppressive (Mazzola 2004; Schlatter et al. 2017; Weller et al. 2002). Several of these soils possess physical or chemical characteristics which effect pathogen suppression directly or indirectly via alteration of a biological constituent of the ecosystem at the expense of the pathogen (Martin and Loper 1999). Although suppressive soils have been reported for many soil pathogens (Spooren et al. 2024; Weller et al. 2002), only a couple have been described for *Pythium* spp. (Martin and Loper 1999).

About 37% of the soil samples taken at various islands of Hawaii were suppressive to sporangial germination of *Pythium splendens*. Percentage of sporangial germination was inversely related to soil pH and near neutral to alkaline soils in Hawaii tended to more suppressive than acid soils (Kobayashi and Ko 1985). Follow-up studies revealed that suppression of *P. splendens* in these soils is due to a combination of high Ca and a high microbial population. Conducive soils could be made suppressive by amendment with Ca and alfalfa meal (Kao and Ko 1986a, b).

Soils suppressive to *Pythium ultimum*, a major cause of pre-emergence damping off on cotton, in the San Joaquin Valley of California tended to have a finer texture (silty clay loams and silty clay soils) than conducive soils (clay loam soils). In addition, suppressive soils had higher concentrations of sodium, sulfate and chloride than conducive soils. Only chloride inhibited the saprophytic potential of *P. ultimum* and this coincides with an increase in the level of colonization by *Pythium oligandrum*, which is more tolerant to chloride. It was shown that *P. oligandrum* is able to outcompete *P. ultimum* for organic matter colonization (Martin and Hancock 1986).

A survey of soil suppressiveness in ten European long-term field experiments using *P. ultimum—Lepidium sativum* (cress) as a model system revealed that suppressiveness was linked mainly to microbial biomass and soil organic carbon quality (labile carbon fractions), but not to total soil organic matter content. The authors concluded that labile carbon, which is the primary energy source for microorganisms, is important for the maintenance of an abundant and active soil microbial community essential for the expression of soil suppressiveness (Bongiorno et al. 2019).

For *P. myriotylum*, suppressiveness to CRRD has only been described in the volcanic andosols of Cameroon (Adiobo et al. 2007). In Cameroon, two main soil types are used for cocoyam production: ferralsols and andosols. Ferralsols or ferralitic soils are weathered soils mainly composed of kaolinite clay, sesquioxides and gibbsite. They have a low nutrient content, are characterized by Al and Mn toxicity, and have a low water retention. Andosols are soils developed from volcanic ash. They are relatively fertile and have a high nutrient content and a medium water retention. In Cameroon cocoyam can be continuously grown on andosols for about 5 years if the plot is from virgin land. Andosols can become progressively conducive after that period because of land use intensity (Adiobo et al. 2007). Soil samples from three andosols and three ferralsols were collected from fields and greenhouse assays were conducted in order to ascertain their soil suppressiveness to the CRRD. Besides revealing that ferralsols possess higher disease severity than andosols, the results of this study indicated that the suppressiveness of andosols correlated with high presence of organic matter, Ca, K, Mg and N. On the other hand, the high content of sand and clay in conducive ferralsols positively correlated with disease severity. Thus, high organic matter and certain physicochemical properties appears to mediate disease suppression of andosols via improvement of soil structure, nutrient content and to some extent, microbial biomass (Adiobo et al. 2007). Oni et al. (2020) compared the *Pseudomonas* and cyclic lipopeptide diversity of the CRRD suppressive Boteva soil in Cameroon with two degraded andosols in Cameroon (Ekona and Maumu) and two CRRD conducive soils in Nigeria (an alfisol from Ado-Ekiti and an ultisol from Umudike). The cocoyam rhizosphere at Boteva was characterized by a diverse *Pseudomonas* population with strains belonging to the *P. fluorescens* and *Pseudomonas putida* group, with the *P. koreensis* subgroup being dominant within the *P. fluorescens* group. In contrast, *P. putida* strains were predominant in conducive soils. The Boteva soil was characterized by *Pseudomonas* strains producing 11 different CLP types, while the putisolvin CLP type, produced by *P. putida* strains was predominant in conducive soils in both Cameroon and Nigeria. *Pseudomonas* species richness and CLP diversity in the various soils correlated with silt, calcium, organic carbon, total nitrogen and ECEC. The Boteva soil appears to have a general suppressiveness to disease, since this type of suppressiveness is community based and driven by microbial competition among soil saprophytes fuelled by the presence of organic carbon (Schlatter et al. 2017). The loss of disease suppressiveness observed in degraded andosols is probably due to the depletion of organic matter (Oni et al. 2020) (Fig. 2.4).

Endophytic actinobacteria were isolated from healthy white cocoyam roots collected from plants grown in andosols in the South West region of Cameroon. Out of a screening of 171 Actinobacteria, four isolates with strong antagonistic activity against *P. myriotylum* and some other pathogens were identified. The four isolates all belonged to the genus *Streptomyces*, with three of the isolates being potential new species. All antagonistic bacteria had the ability to produce protease, chitinase, lipase and β-1,3-glucanase, while two of the isolates had also cellulase and pectinase activity. *Streptomyces* sp. strain PERM2 showed the strongest antagonistic activity against *P. myriotylum* in vitro and significantly increased the root length of tissue-culture derived cocoyam plantlets, probably due to the production of large amounts of indol acetic acid (Djuidje et al. 2022).

5.5 Suppressive Composts

Besides the employment of disease suppressive soils with albeit rare occurrence, together with cultural and biological control methods against CRRD, attempts have been made to explore the use of composts in the suppression of *P. myriotylum* on cocoyam. In several crops, the use of compost as alternative to chemical control and

Fig. 2.4 White cocoyam grown in a disease suppressive volcanic andosol at Boteva in Cameroon

as part of an integrated management strategy against plant pathogens has been well documented (Hoitink and Fahy 1986).

In Cameroon, three compost types (A, B and C) were studied for potential suppressiveness to CRRD (Adiobo 2006). Compost A consisted of 10% cow manure, 30% tree bark, 30% fallen leaves and 30% sugarcane waste. Compost B composed of 10% poultry manure, 35% sugarcane waste and 55% oil palm waste whereas compost C contained 10% cow manure, 35% green grasses and 55% oil palm waste. Various composts were assessed in green house plant experiments using varying levels of *P. myriotylum* inoculum (400, 800 and 1200 propagules/g soil). Compost doses were 10, 20, and 30%, vol/vol. The three composts varied in their disease suppressive capacity. Compost A was most suppressive followed by compost C and then B. Interestingly, chemical properties determination of all composts showed that compost A had both a high percentage of organic carbon (22.80%) and calcium content (76.20 cmol/kg). Microbial analysis of composts further revealed that fluorescent pseudomonads and actinomycetes were higher in compost A than in other compost types. Interestingly, the disease suppressiveness of all compost types were lost after autoclaving and could only be partly recovered following colonization by their resident flora. These results suggest that the disease suppression demonstrated by all three compost types is predominantly biological and due to the activities of diverse microbial groups. Field experiments using afore-mentioned composts A, B and C were further conducted in Cameroon (Adiobo 2006). Three separate cocoyam fields, a degraded andosol located in Ekona and two ferralsols located in Matomb and Akonolinga, were amended with various composts at the rate of 20 t/ha a month before cocoyam planting while a non-amended plot and a fungicide-treated plot were used as controls. The effects of the treatments on plant growth, soil properties, disease incidence and pathogen population among others, were assessed. In all compost amended soils, a significant increase in plant growth and yield was recorded. While total N and organic C were relatively higher, the population of *P. myriotylum* was relatively lower in compost-amended soils. There was a significant reduction of disease incidence in all compost-amended soils although only that recorded in the degraded andosol at Ekona was satisfactory with a disease suppression of 68–70% (Adiobo 2006). These results indicate that disease suppressiveness in degraded andosols can be restored by supplying organic matter, while this is not the case in ferralsols (Fig. 2.5).

In Costa-Rica, effects of several organic composts were tested against CRRD (Campos et al. 2008). Sterile soils were inoculated with *P. myriotylum* isolate CR3 after which various doses of organic amendments were added a week later. Two different experiments were conducted involving different organic compounds. In the first experiment, sterile/non-sterile sugarcane organic amendment were used while in the second, vermicompost derived from either cow manure or coffee bean peels were used. During this study, biocontrol experiments using several organisms including *Trichoderma*, *Bacillus subtilis* and *Streptomyces* were also used. Of all the organic amendments and biocontrol strains employed, vermicompost derived from cow manure displayed the best potential against CRRD. While organic amendments/vermicompost could delay CRRD attack, the microorganisms used in biocontrol were not effective against CRRD.

Fig. 2.5 (**a**) Disease conducive andosol at Ekona; (**b**) Disease conducive andosol at Ekona supplied with compost A; (**c**) Disease conducive ferrasol at Matomb; (**d**) Disease conducive ferralsol at Matomb supplied with compost A

In Cameroon, eight compost types made from four different grass species namely *Tithonia diversifolia*, *Chromolaena odorata*, *Pennisetum purpureum* and *Ageratum conyzoides*, were evaluated for suppressiveness against the root rot disease of cocoyam (Djeugap et al. 2014). In the first treatment combination, each grass was mixed with poultry manure while in a second compost type, the same grasses were homogenized with pig manure. Furthermore, the physicochemical properties (pH-H_2O, CEC, N, K, Na, Mg and Ca) of compost were determined and analyzed. Results of this study showed significant reduction ($P \leq 0.05$) in disease incidence and severity in all compost amended pots than the control at 12 weeks post inoculation. More so, the microbial populations of heterotrophic bacteria and actinomycetes in the different compost types correlated significantly with CRRD incidence and severity, with *Tithonia diversifolia* recording the highest degree of suppressiveness for both manure types. It therefore appears that disease suppression of *P. myriotylum* on cocoyam can be positively influenced by the activities of microbial populations in combination with physicochemical properties of the compost (Djeugap et al. 2014).

In Nigeria, an integrated plant nutrition approach was adopted in the control of CRRD during several field trials at the National Root Crops Research Institute (NRCRI), Umudike (Chukwu and Eteng 2014). Besides the control treatments, four soil amendments were used namely rice mill waste alone, rice mill waste plus poultry manure, NPK alone and lastly, a combination of rice mill waste, NPK and poultry manure. There were significant differences in the number of rotted corms and cormels with the best control displayed by treatments containing three amendments followed by rice mill waste alone, NPK alone and the least control was given by rice mill waste plus poultry manure. The addition of these amendments further improved total yield, soil pH, total N, organic carbon, available P and exchangeable K. Compared to other treatments, the combined use of 5 t/ha rice mill waste, 4 t/ha poultry manure and 400 kg/ha NPK gave the best results for all parameters measured. Thus, integrated plant nutrition via the combined application of NPK together with organic soil amendments, appeared to impact soil health positively, against CRRD.

5.6 *Induction of Resistance*

Induction of disease resistance via the host defense mechanism has been shown to be of practical application in the control of plant diseases (Walters et al. 2014). A practical method is to induce systemic acquired resistance (SAR) through the use of elicitors. One of such elicitors is the Benzo-(1,2,3)-thiadiazole-7-carbothioic S-methyl ester (BTH). BTH, an analog of salicylic acid, has been developed as a crop protection tool. Exogenous treatment of plants with BTH can trigger systemic acquired resistance which plays a central role in disease resistance (Mbouobda et al. 2010). For example, when applied to a plant, BTH can induce the accumulation of pathogenesis related proteins such as chitinases, glucanases and peroxidases with attendant reduction in disease incidence and severity (Oostendorp et al. 2001).

The application of BTH as foliar sprays effectively elicited defense reactions in cocoyam while enhancing the activities of peroxidase, polyphenol oxidase and phenolic compounds (Mbouobda et al. 2010). Nonetheless, since corms are the main propagation material for cocoyam cultivation coupled with its role as a primary inoculum, it appeared important to attempt to subject the corm to inducer treatments. Consequently, maximum protection against CRRD could be obtained during seedling emergence and early growth stages.

In an attempt to induce resistance to CRRD in Cameroon, the application of two concentration levels (15 and 50 mg/L) of BTH on corms was compared with the fungicide Metalm 72 WP (metalaxyl + cuprous oxide) treatment (Oumar et al. 2015). In this study, corms treated with benzothiadiazole (BTH) were planted in the disease-conducive andosol at Ekona inoculated with *P. myriotylum*. BTH treatment significantly reduced CRRD coupled with a dose-dependent effect in comparison with the untreated diseased control. Furthermore, the application of 50 mg/L of BTH was observed to be more effective than the standard fungicide in reducing CRRD. No direct effect of BTH on the mycelium of *P. myriotylum* was observed when *in vitro* tests were conducted indicating that the protective effect of BTH could only be due to its host defence-inducing activity (Oumar et al. 2015).

5.7 Breeding for CRRD Resistance

A review of 50 years of research efforts about cocoyam genetic resources and breeding has recently been published (Lebot et al. 2024). More than three decades ago, a Root and Tubers research project (ROTREP) was funded in Cameroon by the United States Agency for International Development (USAID) in collaboration with the Government of Cameroon. The core objective of the project was to develop cultivars with resistance/tolerance to the CRRD and acceptable agronomic and sociological characteristics (Onokpise et al. 1999). To kick-start this project, over 300 accessions were obtained from three agro-ecological zones in Cameroon, from Equatorial Guinea, Ghana, Gabon, Puerto Rico and Togo. After collecting relevant data such as source of planting material, materials were planted first in the green house and subsequently in the field. Field evaluation included disease incidence, yield and petiole length but success results arising from these efforts are not clear. Financial support for the field maintenance of the collection ended in 1994 and a few years later a substantial part of the collection was lost due to pests and diseases and poor adaptation to edaphoclimatic conditions (Tambong et al. 1997).

In a related study in Cameroon, cocoyam screening for tolerance and resistance to CRRD was conducted for 571 accessions which were obtained from Cameroon and 55 accessions from Ghana and Togo (Pacumbaba et al. 1992a). Interestingly, all cocoyam accessions obtained from Ghana and Togo appeared to be resistant to CRRD whereas only one ROTREP (Cameroon) accession (presumed to be resistant in the greenhouse) and 21 accessions (presumed to be tolerant), survived in the field trials. A limitation of the study was that *P. myriotylum* isolates from Ghana and Togo were not included in this trial. In addition, there appeared to be a possibility that the *P. myriotylum* isolate used in the greenhouse experiments had lost its virulence thereby accounting for the high turnout of resistant accessions obtained under those conditions in comparison to the field trials.

Conventional breeding for resistant cocoyam varieties have been attempted without success mainly due to its strict asexual reproduction. Efforts aimed at crossing white or red cocoyam with the wild, yellow cocoyam which shows resistance, have failed as a result of ploidy incompatibility (Saborio et al. 2004). In Costa Rica mutation induction with gamma irradiation was implemented to generate variability in cocoyam and 6500 plants were screened in the greenhouse in sterile soil inoculated with a pathogenic *Fusarium* strain isolated from diseased cocoyam plants. Seventeen plantlets showed reduced disease symptoms ((Saborio et al. 2004). A similar study was conducted in Ghana (Blay et al. 2004). Of the 48 *X. sagittifolium* accessions tested, one accession, RAX-93-005, was selected for mutation due to its superior morpho-agronomic characters. Corms of this accession were cut into minisets, each bearing an adventitious bud, and subject to a series of preliminary treatments following which sprouts arising from the minisets were exposed to gamma irradiation of variable intensities. The resulting plants were weaned and transplanted into the field. In this study, about 2000 plantlets were generated and multiplied for further work. The mutant population was evaluated in field experiments but the differences between irradiated and non-irradiated plants were not significant. Effects on susceptibility to CRRD were not mentioned (Danquah et al. 2006).

Several studies have investigated the use of plant defense responses as indicators of cocoyam tolerance or susceptibility to the CRRD. The host defense resistance of cocoyams was investigated in Cameroon (Nyochembeng et al. 2007). Additionally, an attempt was made to ascertain if the plant defense responses observed could be linked to cocoyam tolerance or resistance. Four cocoyam accessions were assessed for reaction to CRRD by inoculation of *P. myriotylum* 'Mamu' isolate via two different methods (Nyochembeng et al. 2007). Roots of intact plants in the green house and detached roots in the laboratory were separately inoculated after which assessment was conducted every 2 days up to 8 days post inoculation (dpi). Measurements of electrolyte leakage in detached infected roots, peroxidase activities and isozyme expressions of protein extracts of both healthy and diseased plants revealed that two cocoyam cultivars, RO1054 and RO3015 were tolerant/resistant to CRRD whereas two other isolates were rated as susceptible. Besides, increased peroxidase activity was observed with tolerant RO1054 at 8 dpi. Results of this study further suggest that electrolyte leakage in infected roots can be used to assess tolerance/susceptibility of cocoyams to CRRD while disease tolerance could be correlated with increased peroxidase activity (Nyochembeng et al. 2007).

6 Conclusions and Outlook

The persistence in incidence and spread of CRRD represents a threat to cocoyam production throughout the tropics. In West and Central Africa, it is even more a crucial concern owing to the outbreaks of taro leaf blight caused by *Phytophthora colocasiae* on *Colocasia esculenta* (Bandyopadhyay et al. 2011; Omane et al. 2011) in West Africa (Onyeka 2014). This foliar pathogen causes rapid defoliation of the taro and often kills plants entirely leading to a significantly reduction if not complete eradication of taro production in some regions. Despite considerable efforts, cocoyam cultivars with resistance to CRRD are not yet available (Lebot et al. 2024) and chemical control of the disease is not satisfactory. In absence of resistant cultivars, farmers have to rely on cultural practices to manage this disease. Based on the knowledge obtained after years of research in West Africa an integrated approach can be proposed. At least in West Africa, *Pythium myriotylum* var. *aracearum* is not present in virgin soils, which means that farmers should avoid further spreading of the pathogen via infected planting material. In Cameroon, efforts have been made to produce disease-free cocoyam planting material based on the multiplication of healthy suckers (Fig. 2.6a). To reduce the amount of *P. myriotylum* in already infected soils soil organic amendments based on compost and crop rotation can be used. To improve the plant's ability to escape the disease early planting and soil fertilization is recommended (Fig. 2.6b).

There exists a critical need to establish an *in vitro* collection at world level to aid the evaluation of genetic potential with respect to current needs and problems.

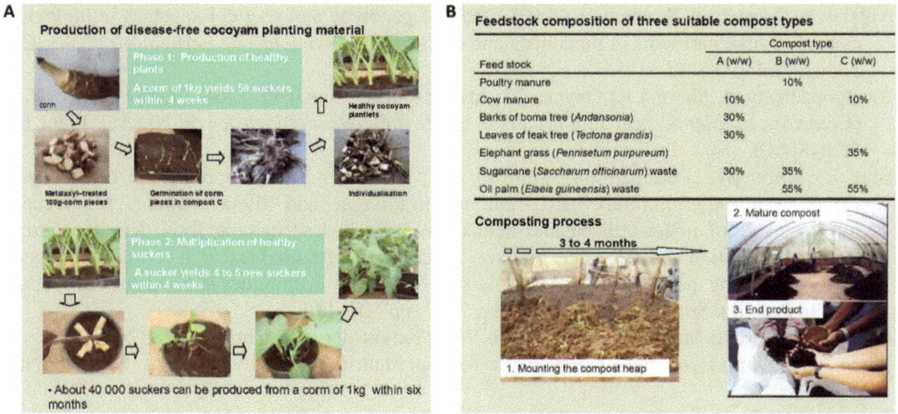

Feed stock	Compost type A (w/w)	B (w/w)	C (w/w)
Poultry manure		10%	
Cow manure	10%		10%
Barks of boma tree (*Andansonia*)	30%		
Leaves of teak tree (*Tectona grandis*)	30%		
Elephant grass (*Pennisetum purpureum*)			35%
Sugarcane (*Saccharum officinarum*) waste	30%	35%	
Oil palm (*Elaeis guineensis*) waste		55%	55%

Fig. 2.6 (**a**) Procedure to produce disease-free cocoyam planting material; (**b**) Procedure to produce disease suppressive compost

Besides collecting known cultivars from cocoyam growing regions across the world, this should further entail the search in the northern part of South America for possible wild forms and primitive cultivars in addition to related species. This collection could then serve as a basis for genetic improvement.

A notable limitation experienced in tannia production is the absence of information or germplasm exchange between producing regions as was recently also pointed out by Lebot et al. (2024). When significant progress is made in a region as it was in Cameroon with the causal pathogen of CRRD for instance, knowledge about this finding did not immediately spread far due to geographical or perhaps linguistic limitations. To correct this, it is therefore necessary to establish a centralized information system such as cooperatives or online networks to which all concerned countries and researchers have open access. This will both enable and enhance *in vitro* germplasm exchange and visits from farmers and scientists thereby aiding knowledge and expertise transfer.

References

Acevedo-Rodriguez P, Strong MT (2012) Catalogue of seed plants of the West Indies. In: Smithsonian contributions to botany, vol 98. Smithsonian Institution, Washington, DC

Adiobo A (2006) Biological control of cocoyam (*Xanthosoma sagitifolium* (L.) Schott) root rot disease caused by *Pythium myriotylum* Dreschl.: importance of soil organic matter content and cultural practices. Ghent University

Adiobo A, Oumar O, Perneel M, Zok S, Höfte M (2007) Variation of *Pythium*-induced cocoyam root rot severity in response to soil type. Soil Biol Biochem 39:2915–2925

Anjaiah V, Koedam N, Nowak-Thompson B, Loper JE, Höfte M, Tambong JT, Cornelis P (1998) Involvement of phenazines and anthranilate in the antagonism of *Pseudomonas aeruginosa* PNA1 and Tn5 derivatives toward Fusarium spp. and Pythium spp. Mol Plant-Microbe Interact 11:847–854

Anjaiah V, Cornelis P, Koedam N (2003) Effect of genotype and root colonization in biological control of fusarium wilts in pigeonpea and chickpea by *Pseudomonas aeruginosa* PNA1. Can J Microbiol 49:85–91

Bandyopadhyay R, Sharma K, Onyeka TJ, Aregbesola A, Kumar PL (2011) First report of Taro (*Colocasia esculenta*) leaf blight caused by *Phytophthora colocasiae* in Nigeria. Plant Dis 95:618–618

Blay ET, Offei SK, Danquah EY, Amoatey HA, Asare E (2004) Improvement of Cocoyam (*Xanthosoma sagittifolium*) using gamma irradiation and tissue culture. In: Genetic Improvement of under-utilised and neglected crops in low income food deficit countries through irradiation and related techniques. Proceedings of a final research coordination meeting organized by the Joint FAO/IAEA Division of Nuclear Techniques in Food and Agriculture held in Pretoria, South Africa, 19-23 May, 2003, pp 127–130

Boakye AA, Wireko-Manu FD, Oduro I, Ellis WO, Gudjónsdóttir M, Chronakis IS (2018) Utilizing cocoyam (*Xanthosoma sagittifolium*) for food and nutrition security: a review. Food Sci Nutr 6:703–713

Bongiorno G, Postma J, Bünemann EK, Brussaard L, de Goede RGM, Mäder P, Tamm L, Thuerig B (2019) Soil suppressiveness to *Pythium ultimum* in ten European long-term field experiments and its relation with soil parameters. Soil Biol Biochem 133:174–187

Boudjeko T, Andème-Onzighi C, Vicré M, Balangé A-P, Ndoumou DO, Driouich A (2006) Loss of pectin is an early event during infection of cocoyam roots by *Pythium myriotylum*. Planta 223:271–282

Boyce PC, Croat TB (2011) Onwards. The Überlist of Araceae, Totals for Published and Estimated Number of Species in Aroid Genera

Brooks FE (2008) Detached-leaf bioassay for evaluating Taro resistance to *Phytophthora colocasiae*. Plant Dis 92:126–131

Brown D (2000) Aroids. Plants of the arum family, 2nd edn. Timber Press, Portland

Campos N, Uribe L, Saborío F (2008) Evaluation of different biological agents on the control of *Pythium myriotylum* attack on cocoyam (*Xanthosoma sagittifolium*) plants. First International Workshop on Cocoyam (*Xanthosoma sagittifolium*) - Cultivation, Genetic Improvement and Disease, Institute of Agricultural Research for Development (IRAD), Ekona, Cameroon

Chin-A-Woeng TFC, Bloemberg GV, Van Der Bij AJ, Van Der Drift KMGM, Schripsema J, Kroon B, Scheffer RJ, Keel C, Bakker PAHM, Tichy HV, De Bruijn FJ, Thomas-Oates JE, Lugtenberg BJJ (1998) Biocontrol by phenazine-1-carboxamide-producing *Pseudomonas chlororaphis* PCL1391 of tomato root rot caused by *Fusarium oxysporum* f. sp. *radicis-lycopersici*. Mol Plant-Microbe Interact 11:1069–1077

Chukwu GO, Eteng EU (2014) Enhancing soil health and control of cocoyam root rot through integrated plant nutrition. J Soil Environ Sci 2:40–45

Cobley LS, Steele WM (1976) An introduction to the botany of tropical crops. Longman

D'aes J, Hua GKH, De Maeyer K, Pannecoucque J, Forrez I, Ongena M, Dietrich LEP, Thomashow LS, Mavrodi DV, Höfte M (2011) Biological control of *Rhizoctonia* root rot on bean by phenazineand cyclic lipopeptide-producing *Pseudomonas* CMR12a. Phytopathology 101:996–1004

D'aes J, Kieu NP, Léclère V, Tokarski C, Olorunleke FE, De Maeyer K, Jacques P, Höfte M, Ongena M (2014) To settle or to move? The interplay between two classes of cyclic lipopeptides in the biocontrol strain *Pseudomonas* CMR12a. Environ Microbiol 16:2282–2300

Danquah A, Offei SK, Blay ET, Asare E, Danquah EY (2006) Characterization of a mutant population of cocoyam (*Xanthosoma sagittifolium* L. Schott). Int J Bot 2:128–132

Djeugap JF, Azia TA, Fontem DA (2014) Effect of compost quality and microbial population density of composts on the Suppressiveness of *Pythium myriotylum*, causal agent of cocoyam (*Xanthosoma sagittifolium*) root rot disease in Cameroon. Int J Sci Basic Appl Res (IJSBAR) 15:209–218

Djuidje PFK, Asultan W, Beaulieu C, Wong MY, Boudjeko T (2022) Characterization of endophytic *Streptomyces* strains from roots of cocoyam (*Xanthosoma sagittifolium* L. Schott) in the

South West Region of Cameroon, their in vitro plant growth promoting abilities and biocontrol efficacy against *Pythium myriotylum*. S Afr J Bot 144:145–155
Drechsler C (1930) Some new species of *Pythium*. J Washington Acad Sci:398–418
Giacometti DC, León J (1994) Tannia, yautia (*Xanthosoma sagittifolium*). Neglected Crops 1492:253–258
Girard L, Lood C, Höfte M, Vandamme P, Rokni-Zadeh H, Van Noort V, Lavigne R, De Mot R (2021) The ever-expanding *pseudomonas* genus: description of 43 new species and partition of the *Pseudomonas putida* Group. Microorganisms 9:1766–1766
Gomes de Souza P, Rosane P, Azeredo D, da Silva TTC, Carneiro CDS, Junger Teodoro A, Menezes Ayres EM (2023) Food neophobia, risk perception and attitudes associations of Brazilian consumers towards non-conventional edible plants and research on sale promotional strategies. Food Res Int 167:112628
Gómez-Alpízar L, Saalau E, Picado I, Tambong JT, Saborío F (2011) A PCR-RFLP assay for identification and detection of *Pythium myriotylum*, causal agent of the cocoyam root rot disease. Lett Appl Microbiol 52:185–192
Govaerts R, Frodin DG, Royal Botanic Gardens, K (2002) World checklist and bibliography of Araceae (and Acoraceae). Royal Botanic Gardens, Kew
Höfte M (2021) The use of pseudomonas Spp. as bacterial biocontrol agents to control plant disease. In: Köhl J, Ravensberg W (eds) Microbial bioprotectants for plant disease management. Burleigh Dodds, pp 301–374
Höfte M, Altier N (2010) Fluorescent pseudomonads as biocontrol agents for sustainable agricultural systems. Res Microbiol 161:464–471
Hoitink HAJ, Fahy PC (1986) Basis for the control of Soilborne plant pathogens with composts. Annu Rev Phytopathol 24:93–114
Kao CW, Ko WH (1986a) Suppression of *Pythium splendens* in a Hawaiian soil by calcium and microorganisms. Phytopathology 76:215–220
Kao CW, Ko WH (1986b) The role of calcium and microorganisms in suppression of cucumber damping-off caused by *Pythium splendens* in a Hawaiian soil. Phytopathology 76:221–225
Kobayashi N, Ko WH (1985) *Pythium splendens*-suppressive soils from Different Islands of Hawaii. Soil Biol Biochem 17:889–891
Lebot V, Ivančič A, Lawac F (2024) Cocoyam (*Xanthosoma sagittifolium* (L.) Schott) genetic resources and breeding: a review of 50 years of research efforts. Genet Resour Crop Evol 72:2593–2612
Ma Z, Geudens N, Kieu NP, Sinnaeve D, Ongena M, Martins JC, Höfte M (2016) Biosynthesis, chemical structure, and structure-activity relationship of orfamide lipopeptides produced by *Pseudomonas protegens* and related species. Front Microbiol 7:382
Martin FN, Hancock JG (1986) Association of chemical and biological factors in soils suppressive to *Pythium ultimum*. Phytopathology 76:1221–1231
Martin FN, Loper JE (1999) Soilborne plant diseases caused by *Pythium* spp.: ecology, epidemiology, and prospects for biological control. Crit Rev Plant Sci 18:111–181
Mazzola M (2004) Assessment and management of soil microbial community structure for disease suppression. Annu Rev Phytopathol 42:35–59
Mbarga JB, Ten Hoopen GM, Kuaté J, Adiobo A, Ngonkeu MEL, Ambang Z, Akoa A, Tondje PR, Begoude BAD (2012) *Trichoderma asperellum*: A potential biocontrol agent for *Pythium myriotylum*, causal agent of cocoyam (*Xanthosoma sagittifolium*) root rot disease in Cameroon. Crop Prot 36:18–22
Mbouobda HD, Fotso, Djocgoue PF, Omokolo ND, El Hadrami I, Boudjeko T (2010) Benzo-(1,2,3)-thiadiazole-7-carbothioic S-methyl ester (BTH) stimulates defense reactions in *Xanthosoma sagittifolium*. Phytoparasitica 38:71–79
Misra RS, Sharma K, Mishra AK (2008) Phytophthora leaf blight of Taro (*Colocasia esculenta*) – a review. Asian Aust J Plant Sci Biotechnol 2:55–63
Moorman GW, Kang S, Geiser DM, Kim SH (2002) Identification and characterization of *Pythium* species associated with greenhouse floral crops in Pennsylvania. Plant Dis 86:1227–1231

Mulet M, Martínez MJ, Gomila M, Dabernig-Heinz J, Wagner GE, Kittinger C, Zarfel G, Lalucat J, García-Valdés E (2023) Genome-based species diversity assessment in the *pseudomonas chlororaphis* phylogenetic subgroup and proposal of *pseudomonas danubii* sp. nov. Isolated from freshwaters, soil, and rhizosphere. Diversity 15:617

Nyochembeng LM, Garton S (1998) Plant regeneration from cocoyam callus derived from shoot tips and petioles. Plant Cell Tiss Org 53:127–134

Nyochembeng LM, Pacumbaba RP, Beyl CA (2002) Calcium enhanced zoospore production of *Pythium myriotylum* in vitro. J Phytopathol 150:396–398

Nyochembeng LM, Beyl CA, Pacumbaba RP (2007) Peroxidase activity, isozyme patterns and electrolyte leakage in roots of cocoyam infected with *Pythium myriotylum*. J Phytopathol 155:454–461

Nzietchueng S (1983) La pourriture racinaire du macabo (*Xanthosoma sagittifolium*) au Cameroun: I—Symptomatologie et étiologie de la maladie. L'Agronomie tropicale 1975(38):321–325

Nzietchueng S (1984) Root rot of Xanthosoma sagittifolium caused by Pythium myriotylum in Cameroon. In: Terry ER, Doku EV, Arene OB, Mahungu NN (eds) Tropical root crops: production and uses in Africa, pp 185–188

Nzietchueng S (1985) Genre *Xanthosoma* (macabo) et contraintes de production: cas particulier de la pourriture racinaire causée par *Pythium myriotylum* Drechsl. au Cameroun. Doctoral thesis, University of Yaoundé

Ofoegbu CAE (1993) Root-rot disease of cocoyam (Xanthosoma-Sagittifolium (L) Schott) in the South Eastern State of Nigeria. Discov Innov 5:373–376

Olorunleke FE, Adiobo A, Onyeka JT, Höfte M (2014) Cocoyam root rot disease caused by *Pythium myriotylum* in Nigeria. In: Tropentag 2014. Czech University of Life Sciences, Prague

Olorunleke FE, Kieu NP, Höfte M, Kieu Phuong N, Höfte M (2015) In: Murillo J, Vinatzer BA, Jackson RW, Arnold DL (eds) Recent advances in *Pseudomonas* biocontrol. Caister Academic Press, pp 167–198

Omane E, Oduro KA, Cornelius EW, Opoku IY, Akrofi AY, Sharma K, Kumar PL, Bandyopadhyay R (2011) First report of leaf blight of Taro (*Colocasia esculenta*) caused by *Phytophthora colocasiae* in Ghana. Plant Dis 96:292–292

Oni FE, Olorunleke OF, Höfte M (2019a) Phenazines and cyclic lipopeptides produced by *Pseudomonas* sp. CMR12a are involved in the biological control of *Pythium myriotylum* on cocoyam *(Xanthosoma sagittifolium)*. Biol Control 129:109–114

Oni FE, Geudens N, Omoboye OO, Bertier L, Hua HGK, Adiobo A, Sinnaeve D, Martins JC, Höfte M (2019b) Fluorescent *Pseudomonas* and cyclic lipopeptide diversity in the rhizosphere of cocoyam (*Xanthosoma sagittifolium)*. Environ Microbiol 21:1019–1034

Oni FE, Geudens N, Onyeka JT, Olorunleke OF, Salami AE, Omoboye OO, Arias AA, Adiobo A, De Neve S, Ongena M, Martins JC, Höfte M (2020) Cyclic lipopeptide-producing *Pseudomonas koreensis* group strains dominate the cocoyam rhizosphere of a *Pythium* root rot suppressive soil contrasting with *P. putida* prominence in conducive soils. Environ Microbiol 22:5137–5155

Onokpise OU, Tambong JT, Nyochembeng L, Wutoh JG (1992) Acclimatization and flower induction of tissue culture derived cocoyam (*Xanthosoma sagittifolium* Schott) plants. Agronomie 12:193–199

Onokpise O, Wutoh J, Ndzana X, Tambong J, Meboka M, Sama E, Nyochembeng L, Aguegia A, Nzietchueng S, Wilson J, Burns M (1999) Evaluation of Macabo cocoyam germplasm in Cameroon. In: Perspectives on new crops and new uses, pp 394–396

Onwueme IC (1978) The tropical tuber crops: yams, cassava, sweet potato, and Cocoyams. Wiley

Onwueme IC, Charles WB (1994) Tropical root and tuber crops: production, perspectives and future prospects. FAO

Onyeka JT (2014) Status of cocoyam (*Colocasia esculenta* and *Xanthosoma* spp) in West and Central Africa: production, household importance and the threat from Leaf Blight. Lima (Peru)

Oostendorp M, Kunz W, Dietrich B, Staub T (2001) Induced disease resistance in plants by chemicals. Eur J Plant Pathol 107:19–28

Oumar D, Akumah BE, Tchinda ND (2015) Induction of resistance in cocoyam (*Xanthosoma sagittifolium*) to *Pythium myriotylum* by corm treatments with Benzothiadiazole and its effect on vegetative growth. J Exp Agric Int 5:164–171

Pacumbaba RP, Wutoh JG, Meboka MMB (1992a) Protocol to screen cocoyam accessions for resistance or tolerance to cocoyam root rot disease in Cameroon. Plant Dis 76:768–770

Pacumbaba RP, Wutoh JG, Eyango SA, Tambong JT, Nyochembeng LM (1992b) Isolation and pathogenicity of rhizosphere fungi of cocoyam in relation to cocoyam root-rot disease. J Phytopathol 135:265–273

Pacumbaba RP, Wutoh JG, Meboka MMB, Tambong JT (1994) Production of zoospores, mode of infection, and inoculum potential of *Pythium myriotylum* propagules on cocoyam. J Phytopathol 140:49–54

Perneel M (2006) The root rot pathogen Pythium myriotylum on cocoyam (Xanthosoma sagittifolium (L.) Schott): intraspecific variability and biological control. Doctoral thesis, Ghent University, Gent

Perneel M, Tambong JT, Adiobo A, Floren C, Saborío F, Lévesque A, Höfte M (2006) Intraspecific variability of *Pythium myriotylum* isolated from cocoyam and other host crops. Mycol Res 110:583–593

Perneel M, Heyrman J, Adiobo A, De Maeyer K, Raaijmakers JM, De Vos P, Höfte M (2007) Characterization of CMR5c and CMR12a, novel fluorescent *Pseudomonas* strains from the cocoyam rhizosphere with biocontrol activity. J Appl Microbiol 103:1007–1020

Perneel M, D'hondt L, De Maeyer K, Adiobo A, Rabaey K, Höfte M (2008) Phenazines and biosurfactants interact in the biological control of soil-borne diseases caused by *Pythium* spp. Environ Microbiol 10:778–788

PIER (2022) Pacific Island ecosystems at risk. Secretariat of the Pacific Regional Environmental Programme

POWO (2025) Plants of the world online. Facilitated by the Royal Botanic Gardens, Kew

PROTA4U (2025) Plant Resources of Tropical Africa

Purseglove JW (1972) Tropical crops: monocotyledons. Halsted Press Division, Wiley

Quero-Garcia J, Lebot V (2010) Taro and Cocoyam. In: Bradshaw JE (ed) Root and tuber crops. Springer, New York, pp 149–172

Ramesh V, John K, Ravindran C, Edison S (2007) Agrotechniques and plant nutrition of tannia (*Xanthosoma* sp.): an overview. J Root Crop 33:1–11

Reddy PP (2015) Tannia, *Xanthosoma sagittifolium*. In: Plant protection in tropical root and tuber crops. Springer India, New Delhi, pp 305–312

Reyes Castro G (2006) Studies on cocoyam (*Xanthosoma* spp.) in Nicaragua, with emphasis on *Dasheen mosaic virus*. PhD Swedish University of Agricultural Sciences, Uppsala

Reyes G, Rönnberg-Wästljung AC, Nyman M (2006) Comparison of field performance between dasheen mosaic virus-free and virus-infected in vitro plants of cocoyam (*Xanthosoma* spp.) in Nicaragua. Exp Agric 42:301–310

Reyes G, Ramsell JNE, Nyman M, Kvarnheden A (2009) Sequence characterization of *Dasheen mosaic virus* isolates from cocoyam in Nicaragua. Arch Virol 154:159–162

Saborio F, Umana G, Solano W, Amador P, Munos G, Valerin AT, Torres S, Valverde R (2004) Induction of genetic variation in *Xanthosoma* spp. In: Genetic improvement of under-utilized and neglected crops in low income food deficit countries through irradiation and related techniques. FAO/IAEA, Pretoria, pp 143–154

Schafer J (1999) Amélioration du système de culture du Macabo, *Xanthosoma sagittifolium* (L.) Schott, en pays Bamiléké (Ouest-Cameroun). Cah Agric 8:9–20

Schlatter D, Kinkel L, Thomashow L, Weller D, Paulitz T (2017) Disease suppressive soils: new insights from the soil microbiome. Phytopathology 107:1284–1297

Spooren J, van Bentum S, Thomashow LS, Pieterse CMJ, Weller DM, Berendsen RL (2024) Plant-driven assembly of disease-suppressive soil microbiomes. Annu Rev Phytopathol 62:1–30

Steiner KG (1981) A root rot of Macabo (*Xanthosoma* sp) in Cameroun, associated with *Pythium myriotylum*. J Plant Dis Protect 88:608–613

Tambong JT, Höfte M (2001) Phenazines are involved in biocontrol of *Pythium myriotylum* on cocoyam by *Pseudomonas aeruginosa* PNA1. Eur J Plant Pathol 107:511–521

Tambong JT, Ndzana X, Wutoh JG, Dadson RB (1997 Variability and germplasm loss in the Cameroon national collection of cocoyam (*Xanthosoma sagittifolium Schott* (L.))

Tambong JT, Poppe J, Höfte M (1999) Pathogenicity, electrophoretic characterisation and in planta detection of the cocoyam root rot disease pathogen, *Pythium myriotylum*. Eur J Plant Pathol 105:597–607

Tambong JT, Perneel M, Höfte M, Lévesque CA (2006) Intraspecific variability within *Pythium*. Phytopathology 96:S146–S146

Tchameni SN, Cotârleţ M, Ghinea IO, Bedine MAB, Sameza ML, Borda D, Bahrim G, Dinică RM (2020) Involvement of lytic enzymes and secondary metabolites produced by *Trichoderma* spp. in the biological control of *Pythium myriotylum*. Int Microbiol 23:179–188

Théberge RL (1985) Common African pests and diseases of cassava, yam, sweet potato and cocoyam. International Institute of Tropical Agriculture

Tojo M, Ono H, Nakashima C, Yoneyama S, Jayakody JAS (2005) First report of root rot of cocoyam caused by *Pythium myriotylum* in Sri Lanka. Plant Dis 89:1132–1132

Tondje PR, Roberts DP, Bon MC, Widmer T, Samuels GJ, Ismaiel A, Begoude AD, Tchana T, Nyemb-Tshomb E, Ndoumbe-Nkeng M, Bateman R, Fontem D, Hebbar KP (2007) Isolation and identification of mycoparasitic isolates of *Trichoderma asperellum* with potential for suppression of black pod disease of cacao in Cameroon. Biol Control 43:202–212

van West P, Appiah AA, Gow NAR (2003) Advances in research on oomycete root pathogens. Physiol Mol Plant Pathol 62:99–113

Walters DR, Newton AC, Lyon GD (2014) Induced resistance for plant defense. Wiley

Waterhouse GM, Waterston JM (1966) Pythium myriotylum. CMI Descriptions of pathogenic fungi and bacteria, p 118

Weller DM, Raaijmakers JM, Gardener BB, Thomashow LS (2002) Microbial populations responsible for specific soil suppressiveness to plant pathogens. Annu Rev Phytopathol 40:309–348

Chapter 3
Biological Control of Verticillium Wilt Using Endophytic *Verticillium* from Disease Suppressive Soils

Shirley Marcou and Monica Höfte

1 Introduction

Soil-borne pathogens pose a significant challenge to global agriculture, causing severe economic losses across various crops. Their persistence in soil makes them particularly challenging to manage using conventional control strategies. Suppressive soils offer a promising avenue for identifying and harnessing effective biocontrol agents. While soil suppressiveness has been extensively studied in relation to Fusarium wilts (*e.g.* Alabouvette 1999; Cha et al. 2016; Scher and Baker 1980), suppressive soils for *Verticillium* spp. remain comparatively underexplored.

The fungal genus *Verticillium* is relatively small, comprising only ten species, but it is agriculturally significant, with members displaying a diversity of fungal lifestyles (Inderbitzin et al. 2011). Among these, *Verticillium dahliae* and *Verticillium albo-atrum* are considered the most notorious due to their broad host range. Conversely, in *Brassica* crops, *Verticillium longisporum*—a hybrid of *V. dahliae*—is the primary pathogen responsible for wilting (Depotter et al. 2016). The persistence of *Verticillium* species is largely attributed to their long-lived survival structures, known as microsclerotia. These black melanized, clumped structures, formed by budding of mycelial cells, along with their adaptation to xylem habitats, make *Verticillium* particularly difficult to control (Yadeta and Thomma 2013).

Two key mechanisms underpinning suppressive soils in the management of Verticillium wilt have been identified: the presence of fungal endophytes and the application of organic amendments. Both mechanisms negatively correlate with Verticillium wilt incidence, likely by enhancing microbial competition, inducing

S. Marcou · M. Höfte (✉)
Faculty of Bioscience Engineering, Ghent University, Gent, Belgium
e-mail: monica.hofte@ugent.be

S. Gnanamanickam, M. Höfte (eds.), *Disease Suppressive Tropical Soils*,
https://doi.org/10.1007/978-3-032-08100-1_3

plant defences, or directly antagonizing the pathogen. This review delves deeper into the role of fungal endophytes in disease suppressiveness, highlighting their potential in mitigating *Verticillium*-induced crop losses.

2 Endophytic *Verticillium tricorpus*-Like Species and Soil Suppressiveness

Within the *Verticillium* genus, certain isolates or species can colonize plant hosts asymptomatically, without causing significant vascular discoloration or external wilt-like symptoms. These are commonly referred to as endophytes (*e.g.* Malcolm et al. 2013; Puri et al. 2021; Qin et al. 2008; Tyvaert et al. 2014). Notably, multiple *Verticillium* species can coexist in the soil, including the pathogenic *V. longisporum* and *V. dahliae*, alongside *V. tricorpus*-like organisms (Davis et al. 2000; Davis and Sorensen 1985; França et al. 2013).

Over the past decade, phylogenetic analyses have redefined the *V. tricorpus* morphospecies into three distinct lineages: *V. isaacii, V. klebahnii*, and *V. tricorpus sensu stricto* (Inderbitzin et al. 2011). Among these, *V. tricorpus* and *V. isaacii* have been associated with soil suppressiveness against Verticillium wilt in potato and cauliflower fields, respectively (Davis et al. 2000; França et al. 2013).

2.1 Suppressiveness in Potato Fields

The suppressive potential of *V. tricorpus* was first documented in potato fields in Idaho, USA, where its abundance correlated negatively with Verticillium wilt incidence over multiple years (Davis et al. 2000; Davis and Sorensen 1985). Comparative analyses of suppressive and non-suppressive soils revealed a consistent 20–40% reduction in wilt incidence in suppressive fields, which harboured significantly higher *V. tricorpus* colony-forming units (CFUs) per gram of soil. In contrast, *V. tricorpus* was nearly undetectable in non-suppressive soils. Although *V. dahliae* populations remained relatively stable across soil types for most of the study period, they were markedly lower in suppressive soils by the final year of investigation. These findings suggest that *V. tricorpus* contributes to disease suppression and may enhance potato yield in affected soils.

Further investigations by Davis et al. (2000) demonstrated that *V. tricorpus* from suppressive soils exhibited endophytic or weakly pathogenic behaviour in potato plants under controlled conditions. *V. tricorpus* conferred protection against *V. dahliae* infection and significantly improved potato yield, reinforcing the hypothesis that *V. tricorpus* contributes to soil suppressiveness (Davis et al. 2000; Davis and Sorensen 1985). More recently, Molina et al. (2023) reported a similar negative correlation between *V. tricorpus* CFU density in soil and both *V. dahliae* detection and disease development in potato fields in Manitoba, Canada, further highlighting its potential as a biological control agent.

2.2 Suppressiveness in Cauliflower Fields

In a cauliflower field located in Ardooie, Belgium (West Flanders), *V. isaacii* was associated with suppressiveness against *V. longisporum*, a major pathogen of cruciferous crops (França et al. 2013) (Fig. 3.1). The study site was divided into two distinct sections: one under continuous cauliflower cultivation for over 20 years and the other subjected to crop rotation with potato (*cv. Première*). Interestingly, continuous cauliflower cultivation did not promote *V. tricorpus*-like populations, whereas crop rotation with potato did. Quantitative real-time PCR analysis revealed that potato cultivation primarily stimulated *V. dahliae*. Given that *V. dahliae* isolates from potato are generally non-pathogenic or exhibit reduced pathogenicity toward cauliflower (Bhat and Subbarao 2001; Debode et al. 2005), it is plausible that this rotation strategy contributed to an environment favouring soil suppressiveness.

An isolate retrieved from the suppressive section, Vt305, showed 100% rDNA ITS sequence identity with *V. isaacii* strain PD343, previously characterized as non-pathogenic to lettuce and artichoke (Inderbitzin et al. 2011). Experimental studies demonstrated that Vt305 significantly reduced *V. longisporum*-induced symptom development and colonization in cauliflower under both greenhouse and field conditions (Deketelaere et al. 2020; Tyvaert et al. 2014). Under controlled conditions, Vt305 colonized cauliflower roots, hypocotyls, and stems asymptomatically. Its

Fig. 3.1 Verticillium wilt caused by *Verticillium longisporum* on cauliflower

biocontrol efficacy was dose-dependent, necessitating a balanced ratio between pathogen load or inoculum and the applied biocontrol agent (Tyvaert et al. 2014).

Field trials further revealed that Vt305 applied as microsclerotia on cauliflower seeds significantly mitigated Verticillium wilt in the cauliflower cultivars Clapton and Clarina but not in cultivar Korlanu (Deketelaere et al. 2020). This cultivar-specific variation may be attributed to differences in susceptibility to *V. longisporum*, as previously observed by Debode et al. (2005). In highly susceptible cauliflower cultivars, achieving substantial disease suppression may be more challenging compared to more tolerant or resistant cultivars. However, since comparable susceptibility to *V. longisporum* was reported across the tested cultivars, differential colonization efficiency of *V. isaacii* Vt305 or cultivar-specific interactions with *V. isaacii* and the pathogen may explain the observed variation.

In contrast, when evaluated against *V. dahliae* in *Capsicum annuum* L. (bell pepper), Vt305 did not exhibit notable biocontrol potential under controlled conditions (Tyvaert 2019). Current investigations are examining the role of *Verticillium* isolate Vt305 in interactions between *V. dahliae* and fiber flax (*Linum usitatissimum* L.). Preliminary results indicate that Vt305 holds potential as a biocontrol agent for managing Verticillium wilt in this crop (Moyse et al. unpublished results). These findings collectively highlight the host-specific nature of Vt305. Intriguingly, Vt305 appears effective against Verticillium wilt in cold-adapted crops (cauliflower, fiber flax) but not in warm-adapted species (pepper). This specificity underscores the need for further research to elucidate the mechanisms underlying its specificity and to optimize its application as a biocontrol agent.

2.3 *Morphological and Phylogenetic Differences*

In both aforementioned case studies, the microsclerotia of *V. dahliae*/*V. longisporum*, and *V. tricorpus*-like organisms were distinguished based on their morphology and distribution on soil-like growth media (Goud et al. 2003). Microsclerotia of *V. dahliae* and *V. longisporum* typically appear in a star-like pattern, are smaller, oval to elongated in shape, and form dense clusters. In contrast, microsclerotia of *V. tricorpus*-like organisms are larger, irregularly shaped, and more widely scattered across the medium (Goud et al. 2003) (Fig. 3.2). In addition to morphological distinctions, these groups differ phylogenetically and belong to separate clades. *V. dahliae* and *V. longisporum* are classified within "Clade Flavnonexudans", characterized by the absence of yellow hyphal pigmentation. Conversely, *V. tricorpus*-like organisms belong to "Clade Flavexudans", distinguished by their production of yellow hyphal pigments (Inderbitzin et al. 2011).

These morphological and phylogenetic differences serve as critical diagnostic tools for distinguishing these organisms in studies on soil suppressiveness and host-pathogen interactions. Understanding these variations is essential for elucidating the ecological roles of these species and their potential applications in biological control strategies.

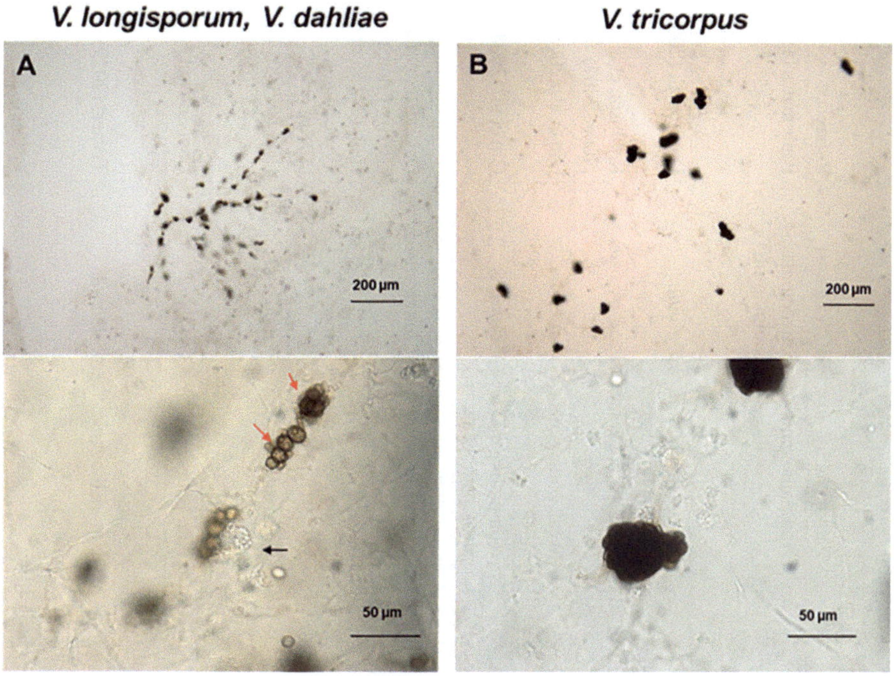

Fig. 3.2 (**a**) Microsclerotia of *Verticillium longisporum* and *V. dahliae*; (**b**) Microsclerotia of *Verticillium tricorpus*

2.4 **V. tricorpus**-*Like Organisms as Biocontrol Agents (BCAs)—Table 3.1*

Already before the disease suppressive characteristics of *V. tricorpus*-like organisms against Verticillium wilt pathogens were fully recognized, early studies explored their potential as BCAs. Matta and Garibaldi (1977) demonstrated that prior inoculation with *V. tricorpus* reduced Verticillium wilt of tomato in greenhouse trials using both naturally infected and artificially inoculated soils, with the degree of control depending on the fungal strain employed, temperature, and soil microbiology. In cotton, Paplomatas and DeVay (1990) achieved cross-protection by treating *V. dahliae*-infested soil with *V. tricorpus*, resulting in a 14% reduction in foliar symptoms following root colonization by *V. tricorpus* (Pegg and Brady 2002).

Building on the trends observed in disease-suppressive soil studies (both in potato and cauliflower) from 2000 onward, subsequent research increasingly focused on employing *V. tricorpus*-like organisms as BCAs.

For example, Robinson et al. (2007) found that *V. tricorpus* reduced potato disease caused by *V. albo-atrum* by 74% in field experiments, decreased colonization of various potato tissues by the pathogen, and that a 4-day pre-inoculation interval

Table 3.1 *Verticillium tricorpus*-like organisms tested as Biocontrol Agents (BCAs) against soil-borne diseases

Host plant	Species old taxonomy	Species new taxonomy	Fungal isolates	Year	Pathogenicity	BCA against	Reference
Gossypium L. (cotton)	*V. tricorpus*	unknown	V-1808/V-15/V-17/V-28/V-31/V-32/V-52/V-54	1991	Non-pathogenic	Rhizoctonia damping-off	Paplomatas (1991)
Solanum tuberosum L. (potato)	*V. tricorpus*	unknown	V-17/V-28/V-31	2000	Non-pathogenic to weak pathogenic	*V. dahliae*	Davis et al. (2000)
Solanum tuberosum L. (potato)	*V. tricorpus*	unknown	not listed	2006/2007	Weak pathogenic	*V. albo-atrum*	Robinson et al. (2006, 2007)
Cynara scolymus L. (artichoke)	*V. tricorpus*	*V. isaacii*	Cs.225/Cs.234/Cs.236/Cs.456	2008	Weak pathogenic	*V. dahliae*	Qin et al. (2008)
Lactuca sativa L. (lettuce)	*V. tricorpus*	*V. isaacii*	Ls. 183/Ls. 432/Ls.441/Ls. 442/Ls. 443	2008	Non-pathogenic to weak pathogenic	*V. dahliae*	Qin et al. (2008)
Brassica oleracea L. var *botrytis* (cauliflower)		*V. isaacii*	Vt305	2013	Non-pathogenic	*V. longisporum*	Deketelaere et al. (2020), França et al. (2013), and Tyvaert et al. (2014)
Lycopersicon esculentum Mill. (tomato)		*V. isaacii*	Vt305/VSO1	2014	Non-pathogenic	*V. dahliae*	Depotter (2014), Tyvaert (2019)
Beta vulgaris L. subsp. *vulgaris* (sugar beet)	*V. tricorpus*	*V. tricorpus*	Vt1808	2017	Non-pathogenic	*V. dahliae*	Lopisso et al. (2017)
Capsicum annuum L. (pepper)		*V. isaacii*	Vt305/VSO1	2018	Non-pathogenic	*V. dahliae*	Tyvaert (2019)

Lycopersicon esculentum Mill. (*tomato*)		*V. klebahnii*	PD659	2021	Non-pathogenic	*V. dahliae*	Puri et al. (2021)
Lycopersicon esculentum Mill. (*tomato*)		*V. isaacii*	PD660	2021	Non-pathogenic	*V. dahliae*	Puri et al. (2021)
Lactuca sativa L. (lettuce)		*V. klebahnii*	PD659	2021	Non-pathogenic	*V. dahliae*	Puri et al. (2021)
Lactuca sativa L. (lettuce)		*V. isaacii*	PD660	2021	Non-pathogenic	*V. dahliae*	Puri et al. (2021)
Solanum melongena L. (eggplant)		*V. klebahnii*	PD659	2021	Non-pathogenic	*V. dahliae*	Puri et al. (2021)
Solanum melongena L. (eggplant)		*V. isaacii*	PD660	2021	Non-pathogenic	*V. dahliae*	Puri et al. (2021)
Linum usitatissimum L. (fiber flax)		*V. isaacii*	Vt305	2025	Non-pathogenic	*V. dahliae*	Moyse et al. (unpublished results)

yielded optimal protection. In lettuce, certain isolates originally identified as *V. tricorpus*—later reclassified as *V. isaacii* (Inderbitzin et al. 2011)—demonstrated potential biocontrol activity against *V. dahliae*, likely through competitive exclusion or cross-protection (Qin et al. 2008). The same authors reported that *V. isaacii* isolates Ls.443, Ls.441, and Ls.432 significantly reduced disease severity caused by *V. dahliae* isolate VdLs.17 (but not VdLs.16), with isolate Ls.443 exhibiting consistent efficacy across inoculation methods; notably, pretreatment provided superior protection compared to co-inoculation.

Given the demonstrated protection of cauliflower against *V. longisporum*, further studies have investigated whether *V. isaacii* can also control *V. dahliae* in other crops, such as tomato and sweet pepper. In organic horticulture in the Netherlands, these solanaceous crops suffer considerable yield losses from Verticillium wilt. In a glasshouse in Brielle, Cuijpers et al. (2015) and Cuijpers et al. (2016) observed that population levels of *V. tricorpus*-like organisms were considerably higher than those of *V. dahliae*, and the *V. isaacii* isolate VSO1 was recovered from the same soil. However, when *V. isaacii* isolates Vt305 (previously shown to suppress Verticillium wilt in cauliflower) and VSO1 were tested against *V. dahliae* strain JR2 in tomato and strain VD1 in pepper using the root-dip method, neither isolate reduced disease symptoms nor colonization. Similarly, in pepper, applications via root-dip or microsclerotia-treated seeds did not decrease symptom development or pathogen colonization. Additional strains (VSO7, VSO19, VSO29, VSO31, and VSO33), selected based on genetic differences in *EF* and *GPD* gene regions, also failed to provide cross-protection; only VSO19 was identified as *V. tricorpus,* with the others confirmed as *V. isaacii* (Tyvaert 2019).

Further studies have expanded the evaluation of non-pathogenic *Verticillium* species as BCAs. Lopisso et al. (2017) tested *V. tricorpus* strain Vt1808 against Verticillium wilt in sugar beet, finding that under controlled conditions, soil application was more effective than seed coating. Notably, a combined application of Vt1808 with a non-pathogenic *Fusarium oxysporum* strain (FoF2) significantly reduced disease incidence and severity on the crown. However, while FoF2 demonstrated promising effects in field trials, the impact of Vt1808 was limited. Additionally, Puri et al. (2021) assessed *V. isaacii* strain PD660 on eggplant, lettuce, and tomato against *V. dahliae* isolate VdLs16 using pre-, post-, and co-inoculation strategies, finding good efficacy in lettuce, intermediate effects in eggplant, and no effect in tomato. In parallel, *V. klebahnii* strain PD659 showed promising control in eggplant and moderate effects in lettuce and tomato.

In addition to non-pathogenic *V. tricorpus*-like organisms, other non-pathogenic *Verticillium* species have also been evaluated for their biocontrol potential. For example, excellent results were obtained with non-pathogenic *V. albo-atrum*—recently revised as *V. alfalfae*, *V. non-alfalfae,* or *V. albo-atrum sensu stricto* (Inderbitzin et al. 2011)—in artificially inoculated tomato soils challenged with *V. dahliae*, although the pre-inoculation effect was more transient in naturally infested soils (Matta and Garibaldi 1977). Moreover, Robinson et al. (2006) reported that a 4-day pre-inoculation with *V. non-alfalfae* (formerly *V. albo-atrum* "group 2") reduced Verticillium wilt symptoms by 65% in field trials. Additional studies in tomato and cotton have further corroborated the biocontrol efficacy of non-pathogenic strains of *V. dahliae* and *V. albo-atrum* (García et al. 2011; Schnathorst and Mathre 1966; Shittu et al. 2009; Zhu et al. 2013).

The mechanisms underlying the antagonistic effects of these *Verticillium* species, as reviewed by Deketelaere et al. (2017), may include plant growth promotion, competition for infection sites, space, and nutrients, and induced resistance through the release of Damage Associated Molecular Patterns (DAMPs). Moreover, Deketelaere et al. (2017) also reviewed other microbial antagonists of Verticillium wilt—such as bacterial, oomycete, and additional fungal strains—which further broadens the scope of potential biocontrol agents against these diseases.

Moreover, *V. tricorpus* has attracted interest for its potential against other soil-borne diseases. For instance, several isolates (V-1808, V-15, V-17, V-28, V-31, V-32, V-52, V-54) have been effective in controlling *Rhizoctonia solani* in cotton under greenhouse conditions, with selected isolates (V-1808; V-15; V-28) showing promising results in field trials against Rhizoctonia damping-off (Paplomatas 1991).

2.5 Drawbacks: Pathogenicity and Diversity Within V. tricorpus*-Like Organisms*

A key limitation of studies evaluating *V. tricorpus*-like species as BCAs is the incomplete understanding of their pathogenic potential across a wide range of horticultural, vegetable, and ornamental crops, as well as their life cycle and ecological interactions in soil. These organisms are frequently described as non-pathogenic or endophytic, exhibiting intermediate saprotrophic ability, yet some isolates have demonstrated mild pathogenicity in certain crops (Barbara and Clewes 2003; Inderbitzin et al. 2011; Inderbitzin and Subbarao 2014; Pegg and Brady 2002). Recently, Wheeler and Johnson (2019) highlighted the variability in fungal lifestyles among *V. isaacii* isolates from a single field, showing that some function as endophytes, while others act as pathogens on potato and sunflower in Washington.

Beyond lifestyle concerns, the complex taxonomic history of *V. tricorpus*-like organisms has introduced uncertainty regarding past data on both their biocontrol efficacy and pathogenicity. Inderbitzin et al. (2011) demonstrated that *V. tricorpus*, as previously mentioned, actually comprises three distinct species: *V. tricorpus*, *V. klebahnii*, and *V. isaacii*. Consequently, isolates used in studies before 2011 likely belong to different species, meaning earlier reports of weakly pathogenic or non-pathogenic *V. tricorpus* isolates may actually refer to *V. isaacii*, *V. klebahnii*, or *V. tricorpus sensu stricto*.

Post-2011 studies remain limited but suggest that *V. tricorpus sensu stricto* is non-pathogenic to sugar beet, with strain Vt1808 exhibiting biocontrol activity against *V. dahliae* under controlled conditions (Lopisso et al. 2017). Research on potato has produced conflicting results, with *V. tricorpus* reported as weakly to highly virulent (Daami-Remadi et al. 2011; Ebihara et al. 2003; Nair et al. 2016, 2019). Conversely, non-pathogenicity has been observed in hosts such as Chinese cabbage, eggplant, green pepper, Larkspur, parsley, snapdragon, soybean, tobacco, and tomato (Ebihara et al. 2003). In okra, pathogenicity has been reported to range from weak to high, whereas in spinach, cotton, tomato, and lettuce, *V. tricorpus* has been described as non-pathogenic (Farag et al. 2024; Iglesias-Garcia et al. 2013). However, it is important to note that the studies by Farag et al. (2024) and Iglesias-Garcia et al. (2013) relied solely on ITS sequences for species determination. These studies did not include all *Verticillium* species in their analysis, and the phylogenetic clustering of isolates with closely related species was not convincingly resolved, leaving room for potential taxonomic misclassification. Remarkably, Usami et al. (2011) reported *V. tricorpus* isolates as highly pathogenic to lettuce, a study that has been widely cited. However, according to NCBI, the Japanese isolates tested—MIH001 (=PD731), Shichi6 (=PD733), Shichi12 (=PD734), and KNty-2-2—are belonging to *V. zaregamsianum*, further complicating interpretations of pathogenicity.

Regarding *V. isaacii*, pathogenicity has been confirmed in several crops, including high virulence on strawberry (Gurung et al. 2015) and pathogenicity on wasabi (Betz and Punja 2021), potato and sunflower (Wheeler and Johnson 2019), artichoke, and some lettuce cultivars (Gurung et al. 2015; Powell et al. 2013). Other studies describe *V. isaacii* as a weak pathogen in artichoke and lettuce (Gurung et al. 2015; Powell et al. 2013; Qin et al. 2008). In contrast, strain Vt305 has been tested via artificial inoculation on cauliflower, potato, chrysanthemum, oilseed rape, broccoli, pepper, fiber flax, tomato, hop, *Phacelia*, and sticky nightshade, without inducing vascular discoloration or Verticillium wilt symptoms (Deketelaere et al. 2020; Depotter 2014; Moyse et al. unpublished results; Tyvaert 2019). Additional *V. isaacii* strains have also been found non-pathogenic in tomato, spinach, pepper, eggplant, cauliflower, cucurbits, *Lagenaria* sp., and certain lettuce cultivars (Depotter 2014; Gurung et al. 2015; Puri et al. 2021; Qin et al. 2008; Tyvaert 2019; Wimer 2015).

In contrast, *V. klebahnii* remains less studied but has been reported as highly virulent on strawberry (Gurung et al. 2015) and pathogenic to artichoke and lettuce (Vallad et al. 2006). Some studies, however, indicate only weak pathogenicity in these crops (Gurung et al. 2015). Non-pathogenicity has been observed in tomato, spinach, eggplant, cauliflower, and certain lettuce cultivars.

These findings highlight the challenges associated with using *V. tricorpus*-like organisms and related species as BCAs, given the variability in their pathogenic potential and the uncertainty surrounding past taxonomic classifications. Further studies incorporating robust phylogenetic analyses and host range assessments are needed to clarify their suitability as BCAs.

3 Conclusions

The few known examples of soils with diseases suppressiveness to Verticillium wilt are all enriched with *Verticillium tricorpus/isaacii* like endophytes with biocontrol potential against pathogenic *Verticillium* spp. These endophytic *Verticillium* spp. share the same niche with pathogenic *Verticillium* spp. where they probably compete for place and nutrients or induce resistance responses in the plant. These organisms can easily be formulated and applied as microsclerotia with a long shelf life. Not all isolates work on all crops, however, and some *V. tricorpus/isaacii* isolates can be pathogenic on specific crops. The challenge will be to select the right isolates and to distinguish between beneficial and potential pathogenic isolates.

References

Alabouvette C (1999) Fusarium wilt suppressive soils: an example of disease-suppressive soils. Australas Plant Pathol 28:57–64. https://doi.org/10.1071/AP99008

Barbara DJ, Clewes E (2003) Plant pathogenic *Verticillium* species: how many of them are there? Mol Plant Pathol 4(4):297–305. https://doi.org/10.1046/j.1364-3703.2003.00172.x

Betz EC, Punja ZK (2021) First report of *Verticillium isaacii* causing wilting and vascular blackening on wasabi (*Wasabia japonica*) plants. Can J Plant Pathol 43(5):658–669. https://doi.org/10.1080/07060661.2020.1869831

Bhat RG, Subbarao KV (2001) Reaction of broccoli to isolates of *Verticillium dahliae* from various hosts. Plant Dis 85(2):141–146. https://doi.org/10.1094/PDIS.2001.85.2.141

Cha J-Y, Han S, Hong H-J, Cho H, Kim D, Kwon Y, Kwon S-K, Crüsemann M, Bok Lee Y, Kim JF, Giaever G, Nislow C, Moore BS, Thomashow LS, Weller DM, Kwak Y-S (2016) Microbial and biochemical basis of a Fusarium wilt-suppressive soil. ISME J 10(1):119–129. https://doi.org/10.1038/ismej.2015.95

Cuijpers W, França S, Debode J, Hospers-Brands A (2015) Stuurbaar bodemleven? Effect van lignine-rijke gewasresten op *Verticillium*. BioKennis 42:1–41

Cuijpers W, Hospers-Brands M, Debode J, França S (2016) Effect of lignin-rich crop residues on the viability of *Verticillium* in organic greenhouse soils. In: III international symposium on organic greenhouse horticulture 1164,

Daami-Remadi M, Jabnoun-Khiareddine H, Ayed F, El Mahjoub M (2011) Comparative aggressiveness of *Verticillium dahliae*, *V. albo-atrum* and *V. tricorpus* on potato as measured by their effects on wilt severity, plant growth and subsequent yield loss. Funct Plant Sci Biotechnol 5:1–8

Davis J, Sorensen L (1985) Association of *Verticillium tricorpus* with *Verticillium dahliae* and potato early dying. Phytopathology 75(11):1279

Davis J, Everson D, Sorensen L, Schneider A (2000) Association of *Verticillium tricorpus* with soil suppressiveness of Verticillium wilt of potato. In: Advances in *Verticillium* research and disease management. American Phytopathological Society, St. Paul, pp 347–351

Debode J, Declercq B, Höfte M (2005) Identification of cauliflower cultivars that differ in susceptibility to *Verticillium longisporum* using different inoculation methods. J Phytopathol 153(5):257–263. https://doi.org/10.1111/j.1439-0434.2005.00965.x

Deketelaere S, Tyvaert L, Franca SC, Hofte M (2017) Desirable traits of a good biocontrol agent against Verticillium Wilt. Front Microbiol 8:1186. https://doi.org/10.3389/fmicb.2017.01186

Deketelaere S, Spiessens K, Pollet S, Tyvaert L, Rooster L, Callens D, Franca SC, Hofte M (2020) Towards practical application of *Verticillium isaacii* Vt305 to control verticillium wilt of cauliflower: exploring complementary biocontrol strategies. Plants 9(11). https://doi.org/10.3390/plants9111469

Depotter JRL (2014) Interacties tussen goedaardige en pathogene Verticillium isolaten in tomaat. Master thesis, Ghent University. Ghent, Belgium.

Depotter JR, Deketelaere S, Inderbitzin P, Tiedemann AV, Hofte M, Subbarao KV, Wood TA, Thomma BPJ (2016) *Verticillium longisporum*, the invisible threat to oilseed rape and other brassicaceous plant hosts. Mol Plant Pathol 17(7):1004–1016. https://doi.org/10.1111/mpp.12350

Ebihara Y, Uematsu S, Nagao H, Moriwaki J, Kimishima E (2003) First report of *Verticillium tricorpus* isolated from potato tubers in Japan. Mycoscience 44(6):481–488. https://doi.org/10.1007/s10267-003-0142-z

Farag FM, Arafa RA, Abou-Zeid MA, Aloufi AS, Abd El Moneim D, Ghebrial EWR (2024) First record of *Verticillium tricorpus* as a causal agent of Verticillium wilt disease in Okra. J Plant Dis Protect 131(2):557–569. https://doi.org/10.1007/s41348-023-00843-4

França SC, Spiessens K, Pollet S, Debode J, De Rooster L, Callens D, Höfte M (2013) Population dynamics of *Verticillium* species in cauliflower fields: Influence of crop rotation, debris removal and ryegrass incorporation. Crop Prot 54:134–141. https://doi.org/10.1016/j.cropro.2013.08.008

García M, Arriagada C, García-Romera I, Ocampo J (2011) Are plant cell wall hydrolysing enzymes of saprobe fungi implicated in the biological control of the *Verticillium dahliae* pathogenesis? Crop Prot 30(1):85–87. https://doi.org/10.1016/j.cropro.2010.09.007

Goud JK, Termorshuizen AJ, Gams W (2003) Morphology of *Verticillium dahliae* and *V. tricorpus* on semi-selective media used for the detection of *V. dahliae* in soil. Mycol Res 107:822–830. https://doi.org/10.1017/s0953756203008050

Gurung S, Short DPG, Hu X, Sandoya GV, Hayes RJ, Koike ST, Subbarao KV (2015) Host range of *Verticillium isaacii* and *Verticillium klebahnii* from artichoke, spinach, and lettuce. Plant Dis 99(7):933–938. https://doi.org/10.1094/PDIS-12-14-1307-RE

Iglesias-Garcia AM, Villarroel-Zeballos MI, Feng C, Toit LJ d, Correll JC (2013) Pathogenicity, virulence, and vegetative compatibility grouping of *Verticillium* isolates from spinach seed. Plant Dis 97(11):1457–1469. https://doi.org/10.1094/PDIS-01-13-0016-RE

Inderbitzin P, Subbarao KV (2014) *Verticillium* systematics and evolution: how confusion impedes Verticillium wilt management and how to resolve it. Phytopathology 104(6):564–574. https://doi.org/10.1094/PHYTO-11-13-0315-IA

Inderbitzin P, Bostock RM, Davis RM, Usami T, Platt HW, Subbarao KV (2011) Phylogenetics and taxonomy of the fungal vascular wilt pathogen *Verticillium*, with the descriptions of five new species. PloS One 6(12):e28341. https://doi.org/10.1371/journal.pone.0028341

Lopisso DT, Kühlmann V, Siebold M (2017) Potential of soil-derived fungal biocontrol agents applied as a soil amendment and a seed coating to control Verticillium wilt of sugar beet. Biocontrol Sci Technol 27(9):1019–1037. https://doi.org/10.1080/09583157.2017.1357800

Malcolm GM, Kuldau GA, Gugino BK, Jiménez-Gasco M, d. M. (2013) Hidden host plant associations of soilborne fungal pathogens: an ecological perspective. Phytopathology 103(6):538–544. https://doi.org/10.1094/PHYTO-08-12-0192-LE

Matta A, Garibaldi A (1977) Control of Verticillium wilt of tomato by preinoculation with avirulent fungi. Netherlands J Plant Pathol 83:457–462. https://doi.org/10.1007/BF03041463

Molina OI, Henriquez MA, Tenuta M (2023) Culture-based determination of *Verticillium* densities in soil overestimates disease pressure of Verticillium wilt of potato in Manitoba. Am J Potato Res 100(4):324–339. https://doi.org/10.1007/s12230-023-09922-6

Nair PVR, Wiechel TJ, Crump NS, Taylor PWJ (2016) Role of *Verticillium dahliae* and *V. tricorpus* naturally infected tubers in causing Verticillium wilt disease, contribution of soil pathogen inoculum and subsequent progeny tuber infection. Australas Plant Pathol 45(5):517–525. https://doi.org/10.1007/s13313-016-0442-3

Nair PVR, Wiechel TJ, Crump NS, Taylor PWJ (2019) Seed tuber incidence, identification and pathogenicity of *Verticillium* species infecting potatoes in South East Australia. Australasian Plant Pathol 48(6):637–650. https://doi.org/10.1007/s13313-019-00667-7

Paplomatas E (1991) Biocontrol of Rhizoctonia damping-off of cotton seedlings by Verticillium tricorpus. University of California, Davis

Paplomatas EE, DeVay JE (1990) Biological control activity of *Verticillium tricorpus* against seedling diseases and Verticillium wilt of cotton. In: Proceedings of the fifth international Verticillium symposium, Leningrad

Pegg GF, Brady BL (2002) Verticillium wilts. CABI Publishing, Wallingford

Powell M, Gundersen B, Miles C, Coats K, Inglis D (2013) First report of Verticillium wilt on lettuce (*Lactuca sativa*) in Washington caused by *Verticillium tricorpus*. Plant Dis 97(7):996–996. https://doi.org/10.1094/PDIS-12-12-1166-PDN

Puri KD, Hu X, Gurung S, Short DPG, Sandoya GV, Schild M, Zhang Y, Zhao J, Anchieta AG, Klosterman SJ, Subbarao KV (2021) *Verticillium klebahnii* and *V. isaacii* isolates exhibit host-dependent biological control of Verticillium wilt caused by *V. dahliae*. PhytoFrontiers 1(4):276–290. https://doi.org/10.1094/phytofr-01-21-0001-r

Qin Q-M, Vallad GE, Subbarao KV (2008) Characterization of *Verticillium dahliae* and *V. tricorpus* isolates from lettuce and artichoke. Plant Dis 92(1):69–77. https://doi.org/10.1094/PDIS-92-1-0069

Robinson N, Platt HW, Hale L (2006) Potato plant and tuber infection and soil colonization by *Verticillium tricorpus* and *Verticillium albo-atrum* "group 2". Can J Plant Pathol 28(4):540–547. https://doi.org/10.1080/07060660609507332

Robinson N, Platt H, Hale LR (2007) Interactions of various *Verticillium* species in combination with *V. albo-atrum* on Verticillium wilt disease development in potato. Am J Potato Res 84:133–141. https://doi.org/10.1007/BF02987136

Scher FM, Baker R (1980) Mechanism of biological control in a *Fusarium*-suppressive soil. Phytopathology 70(5):412–417. https://doi.org/10.1094/Phyto-70-412

Schnathorst W, Mathre D (1966) Cross-protection in cotton with strains of *Verticillium albo-atrum*. Phytopathology 56:1204–1209

Shittu HO, Castroverde DC, Nazar RN, Robb J (2009) Plant-endophyte interplay protects tomato against a virulent *Verticillium*. Planta 229(2):415–426. https://doi.org/10.1007/s00425-008-0840-z

Tyvaert L (2019) Potential of Verticillium endophytes to control Verticillium wilt and insights in the role of ethylene in the plant-fungus interaction. PhD thesis, Ghent University

Tyvaert L, Franca SC, Debode J, Hofte M (2014) The endophyte *Verticillium* Vt305 protects cauliflower against Verticillium wilt. J Appl Microbiol 116(6):1563–1571. https://doi.org/10.1111/jam.12481

Usami T, Kanto T, Inderbitzin P, Itoh M, Kisaki G, Ebihara Y, Suda W, Amemiya Y, Subbarao KV (2011) *Verticillium tricorpus* causing lettuce wilt in Japan differs genetically from California lettuce isolates. J General Plant Pathol 77(1):17–23. https://doi.org/10.1007/s10327-010-0282-x

Vallad GE, Qin Q-M, Grube R, Hayes RJ, Subbarao KV (2006) Characterization of race-specific interactions among isolates of *Verticillium dahliae* pathogenic on lettuce. Phytopathology 96(12):1380–1387. https://doi.org/10.1094/PHYTO-96-1380

Wheeler DL, Johnson DA (2019) *Verticillium isaacii* is a pathogen and endophyte of potato and sunflower in the Columbia Basin of Washington. *Plant Dis 103*(12):3150–3153. https://doi.org/10.1094/PDIS-04-19-0779-RE

Wimer JA (2015) Grafting watermelon to manage Verticillium wilt in Washington state. Washington State University, Washington

Yadeta KA, Thomma BP (2013) The xylem as battleground for plant hosts and vascular wilt pathogens. Front Plant Sci 4:97. https://doi.org/10.3389/fpls.2013.00097

Zhu HQ, Feng ZL, Li ZF, Shi YQ, Zhao LH, Yang JR (2013) Characterization of two fungal isolates from cotton and evaluation of their potential for biocontrol of Verticillium wilt of cotton. J Phytopathol 161(2):70–77. https://doi.org/10.1111/jph.12027

Chapter 4
Banana Disease Suppressive Soils and Mitigation of Panama Disease in China

Zongzhuan Shen, Beibei Wang, Yannan Ou, Shanshan Liu, Chengyuan Tao, Sai Guo, Xuhui Deng, Hongjun Liu, Rong Li, and Qirong Shen

1 Introduction

Banana, an herbaceous monocotyledonous evergreen plant of the genus of *Musa* and *Musaceae* family, is not only an important fruit and cash crop globally, but also ranks as the fourth largest food crop in the world (Heslop-Harrison and Schwarzacher 2007; Dita et al. 2011). Banana has an over 2000-year cultivation history in China, and is mainly cropped in the provinces of Guangdong, Guangxi, Fujian, Taiwan, Yunnan and Hainan, with a small amount of cultivation in Guizhou, Sichuan and Chongqing area. The planted area and total output of banana in China are rising year by year, acting as the global second largest banana producer after India. Although a range of banana cultivars has been found in China, the 'Cavendish Brazilian' variety (AAA) is the most popular and commercially grown variety of banana in China.

Nevertheless, Cavendish banana production has been severely hampered by Panama disease during the past decades in China, a soil-borne wilt disease caused by the fungus *Fusarium oxysporum* f. sp. cubense race 4 (Foc4) (Butler 2013; Ploetz 2006). The typical wilt symptoms of Panama disease include the leaf yellowing, pseudostem splitting, dark brown discoloration of vascular tissues, and plant death (Jeger et al. 1995). In 1996, Panama disease caused by Foc4 was first reported in Guangzhou, Guangdong Province. Subsequently, this disease rapidly spread to the

Z. Shen · Y. Ou · S. Liu · C. Tao · S. Guo · X. Deng · H. Liu · R. Li (✉) · Q. Shen
The Sanya Institute of the Nanjing Agricultural University, Key Lab of Organic-Based Fertilizers of China, Jiangsu Provincial Key Lab for Solid Organic Waste Utilization, Jiangsu Collaborative Innovation Center of Solid Organic Wastes, Educational Ministry Engineering Center of Resource-Saving Fertilizers, College of Resources and Environmental Sciences, Nanjing Agricultural University, Nanjing, Jiangsu, China
e-mail: lirong@njau.edu.cn

B. Wang
School of Tropical Agriculture and Forestry, Hainan University, Haikou, Hainan, China

S. Gnanamanickam, M. Höfte (eds.), *Disease Suppressive Tropical Soils*,
https://doi.org/10.1007/978-3-032-08100-1_4

whole banana production area. For instance, Panama disease was then detected in Zhangzhou, Fujian Province in 2000, a range adjacent to Guangdong Province. The disease subsequently appeared in Hainan Province (2001), Yunnan Province (2009), and Guangxi Province (2012), during the past 20 years. Currently, banana wilt disease has already been reported in all of the main banana producing areas, causing a 30% reduction of banana yield and resulting in vast economic losses in China.

Although the phenomenon of soil suppressiveness to cotton *Fusarium* wilt disease was found as early as 1892, the concept relating to disease-suppressive soil was first proposed in the 1970s (Cook and Rovira 1976; Weller et al. 2002). It was widely accepted that **disease-suppressive soils** refer to soils in which the pathogen cannot establish or persist, or establishes but causes little or no disease, or establishes and causes disease at first, but the disease gradually declines with successive cropping of a susceptible host, even though the pathogen may still persist in the soil. In contrast, non-disease-suppressive soils (also known as **disease-conducive soils**) refer to soils that that are prone to causing diseases. Currently, disease-suppressive soils are prevalent in agricultural systems and have been reported for a variety of crops, *i.e.*, wheat, sugar beet, potatoes, and strawberries.

Depending on their mechanism, disease suppressive soils can be divided as general disease suppressive soils and specific disease suppressive soils. **General suppression** usually refers to the ability of soils to inhibit the growth and activity of soil-borne pathogens to some extent, which originates from the collective competitive and antagonistic activity of the soil microbiome when competing with the pathogens (Weller et al. 2002). General suppression is a natural and preexisting characteristic of soil, which is often effective against a broad spectrum of soilborne diseases and is not transferrable from field to field or soil to soil with very small amounts of microbial inoculum or soil (Janvier et al. 2007). In contrast, **specific suppression** usually has high-efficiency and results from an individual species or select group of microorganisms. Unlike the general suppression, the disease suppression can be gradually transformed into disease-suppressive soil when a certain percentage of specific disease-suppressive soil is inoculated into disease-conducive soil (Westphal and Becker 2000).

Various management strategies, such as biofertilizer application, soil fumigation, rotation, and intercropping, have been proven to be effective in controlling banana Panama disease in China. Recently, plant root-associated microbiomes are usually known as key contributors to soil-borne pathogen suppression and have become a promising way for overcoming wilt disease (Berendsen et al. 2012). It is widely accepted that the study of disease-suppressive soil could provide a novel strategy by using harnessing soil microbiome to explore sustainable alternate managements to suppress pathogens. Therefore, soils suppressive to soil-borne diseases affords the best case for exhibiting that microbes participate in the soil disease suppression to pathogens for support crop health. As the need to reduce chemical pesticide use becomes more widely recognized, disease suppressive soil has emerged as an important implication in the development of green and sustainable agricultural products. Therefore, a rapidly growing body of literature has recently been focusing on deciphering the soil microbiome and manipulating of soil microbiomes to enhance soil disease suppression ability.

On the global scale, banana soil suppressive to Panama disease was first reported in Australia in 1999 (Peng et al. 1999). In China, however, banana Panama disease-suppressive soil was first reported in Hainan area in 2014 (Shen et al. 2015). Subsequently, deciphering the microbial base of soils suppressive to banana Panama disease and manipulating soil microbiome to enhance soil suppressiveness have become major research focuses in China. This chapter aims to introduce the current research progress of soils suppressive to banana Panama disease. It seeks to improve the understanding towards the banana wilt disease suppressive soils and explore possible and effective ways of management to enhance soil suppressive ability in controlling banana Panama disease, ultimately supporting sustainable banana production.

2 Soil Nutrient Matters in Soils Suppressive to Banana Panama Disease

Since microorganisms were first proposed as one of the bases of disease-suppressive soils in 1931 (Henry 1931), a growing body of l studies have gradually confirmed that microbes are involved in the formation and functional preserving of disease-suppressive soils by means of soil heating and sterilization, use of fungicides, and soil transplantation. Although the physicochemical properties of disease-suppressive soils, such as pH, organic matter content, and clay content, can act directly on pathogens to enhance the disease-suppressive ability of soils, they can also act indirectly by influencing the assembly of soil microorganisms to increase their disease suppression ability. Hence, the formation and maintenance of disease-suppressive soils are determined by a combination of biotic and abiotic factors in the soil.

A comparison of soil physicochemical properties between soils suppressive and conducive to banana Panama disease based on one site located in Hainan, China was conducted by Cao et al. (2024). An increase in available soil phosphorus (AP) content was found to be a significant characteristic of disease-suppressive soils, while the soil pH, total phosphorus, total nitrogen, and total carbon levels were also higher in the banana soils suppressive to Panama disease compared to the soils conducive to that disease. In addition, ratios of lower total carbon to available phosphorus (TC/AP) and total nitrogen to available phosphorus (TN/AP), along with a higher ratio of available phosphorus/available potassium (AP/AK), were then found to be beneficial for wilt disease suppression based on soil stoichiometric analyses. This suggests that the relative increase of available phosphorus compared with carbon, nitrogen, and potassium aided in disease control, thus emphasizing the roles of phosphorus deficiency and balanced nutrient availability in soil-borne disease management. Besides, soil nitrogen content was reported as another vital factor involving in suppression. Worth to mention, the form of nitrogen was also associated with banana *Fusarium* wilt suppression since nitrate nitrogen (NO_3^--N), but not the ammonium nitrogen (NH_4^+-N), was associated with disease suppression ability.

Another study based on multiple sites reported by Shen et al. (2022) showed the potential of disease suppressive soils from six geographically distributed banana orchards which differed in their edaphic factors. In detail, a higher value of soil pH and contents of available phosphorus, available potassium, total carbon, and total nitrogen on average in soils suppressive to Panama disease was observed compared to conducive soils.

The improvement of soil nutrient conditions enhances plant defense functions and also supports a robust soil microbial diversity, which serves as a dominate driver in maintaining soil disease-suppressive capacity. For instance, Cao et al. (2024) also revealed through metagenomic analysis that microbial genes associated with disease suppression, *i.e.*, non-ribosomal peptide synthase genes (NRPS), type I polyketide synthase genes (T1PKS), and bacteriocin genes, exhibited higher abundances in disease-suppressive rhizosphere soils compared to disease-conducive soils. Furthermore, Boruta method analysis demonstrated that the abundance of these genes was positively correlated to the contents of soil available phosphorus. This finding suggests that soil available phosphorus plays a critical role in sustaining disease suppression of soil microbiome.

In summary, soils with higher contents in phosphorus, nitrogen and carbon were found to be beneficial to reduce banana Panama disease in South China. Moreover, antagonistic genes of rhizosphere microbiome involving in disease suppression was stimulated for Panama disease suppressive soils, and which may be caused by the enrichment of available phosphorus content in soil.

3 Microbial Base of Banana Suppressive Soils Against Panama Disease

3.1 Bacterial Communities in Banana Suppressive Soils Against Panama Disease

The natural disease suppression ability of agricultural soils has been observed in a range of cropping systems, such as potato soils suppressive to common scab (Meng et al. 2012), wheat soils suppressive to take-all disease (Cook and Rovira 1976), and melon soils suppressive to *Fusarium* wilt disease (Peng et al. 1999). However, limited previous researches have conducted on banana soils suppressive to *Fusarium* wilt disease, not to mention the works focusing on soil microbiome at a more comprehensive level. The soil microbial composition from the two banana plantations, the one showing strong disease suppression ability (potential disease-suppressive soil) after long-term monoculture and another adjacent to disease-suppressive soils exhibiting a *Fusarium* wilt disease incidence (disease-conducive soils) were deciphered by using deep 16S DNA barcode pyrosequencing until 2015 (Shen et al. 2015).

It was reported that banana soils suppressive to Panama disease exhibited distinct bacterial community structures and compositions compared to disease-conducive soils by Shen et al. (2015). Disease-suppressive soils showed higher relative abundances of certain genera within the phylum of Acidobacteria (*e.g.*, *Chthonomonas*, *Pseudomonas*, and *Tumebacillus*) and lower abundances of Bacteroidetes. Simultaneously, the enrichment of *Pseudomonas* was correlated with changes in soil physicochemical properties and available phosphorus content, demonstrating a positive association with disease-suppressive capacity. Although unraveling the precise mechanisms through what microbial community composition or which specific microorganisms enhancing disease-suppressive capacity remains complex, this study revealed that banana soils resistant to Panama disease typically exhibited unique communities with higher microbial diversity and specific microbial taxa. All these microbial indices were identified as potential biomarkers for evaluating soil disease-suppressive capabilities.

3.2 Core Microbial Populations for Geographically Separated Banana Soils Suppressive to Panama Disease

Shen et al. (2015) previously investigated the microbial community composition in banana soils suppressive to Panama disease within a single geographic region. However, even under identical cropping systems, agricultural soil properties may vary significantly across different geographic areas, leading to significant differences in soil microbial community composition. Consequently, current understanding of whether disease-suppressive soil characteristics exhibit consistency across geographically isolated farmlands remains limited. To expand knowledge on universal microbial features of disease-suppressive soils, Shen et al. (2022) conducted a study on soil microbiomes from six geographically separated banana orchards under long-term monoculture in Hainan Province, China. Each study site included two orchards: one banana plantation showing potential Panama disease suppression and another exhibiting susceptibility to the disease. To explore geographically distributed Panama disease-suppressive soils, this study employed both DNA-based high-throughput sequencing and traditional culture-dependent methods. That is, general features of geographically located soils suppressive to wilt disease were proposed to explore keynote microbial groups potentially contributing to disease suppression in the first, and subsequent cultivation-based microcosm experiments were used to assess the potential of some of these taxa to suppress the pathogen.

By investigating six distinct agricultural sites in China (each containing paired orchards with either Panama disease-suppressive or susceptible characteristics), Shen et al. (2022) identified stable microbiome features associated with disease-suppressive soils. The study revealed *Myxococcales*, *Pseudomonadales*, and *Xanthomonadales* was significantly enriched in soils suppressive to banana Panam disease. Additionally, metagenomic analysis of a pair of adjacent orchards

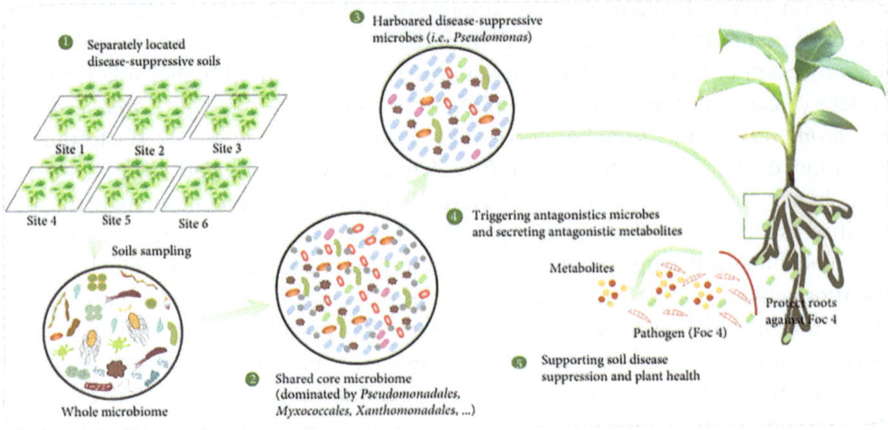

Fig. 4.1 Conceptual figure summarizes the features of core microbiome in separately located disease-suppressive soils and potential mechanism of disease-suppressive microbes against fungal pathogen growth and subsequent plant infection. (Shen et al. 2022)

demonstrated that suppressive soils were enriched with functional genes linked to reduced fungal virulence (*e.g.*, genes encoding enzymes for degrading fungal cell walls or toxins), quorum-sensing genes (a bacterial communication mechanism coordinating antimicrobial synthesis), biofilm formation genes (microbial aggregates protecting plant roots), and antimicrobial compound synthesis genes (directly suppressing pathogens). Furthermore, Shen et al. (2022) reported that culturable *Pseudomonas* strains isolated from suppressive soils exhibited antagonistic to the *Fusarium* pathogen, highlighting their central role in disease suppression. Despite variations in geographic location, soil physicochemical properties, and microbial composition across banana cultivation sites, shared microbiome features associated with disease suppression could serve as universal markers indicative of soil suppressiveness (Fig. 4.1). This study provides a critical theoretical foundation for leveraging soil microbiomes to develop ecological control strategies targeting banana Panama disease.

4 Current Management Strategies in Controlling Banana Panama Disease

It is widely known that Banana Panama disease is relatively difficult to control. Effective prevention and control of Panama disease should adhere to the core concept of "prevention as the mainstay, comprehensive prevention and control", and adopt comprehensive measures such as selecting and breeding disease-resistant varieties, cultivating sterile seedlings, chemical prevention and control, agricultural prevention and control, and biological prevention and control.

4.1 Disease-Resistant Varieties

Breeding of disease-resistant varieties and production of disease-free cultured seedlings are currently recognized as the most effective measures for banana wilt control. Cultivated bananas are mostly triploid, highly sterile and seedless, making conventional crossbreeding of bananas very difficult. At present, the selection and breeding of banana disease-resistant varieties are mainly based on seed introduction, bud mutation, mutation of cultured seedlings, artificial mutagenesis and hybridization. In China, seeds are usually introduced from abroad and then germination and other selective breeding work is carried out, and only a few breeding works has been carried out using seeds of local origin. Through the screening of mutant strains, numerous new banana varieties have been obtained in China, such as 'Baodaojiao', 'Nongke No. 1', 'Kangku No. 5', 'Guijiao No. 9', 'Zhongjiao No. 9' and 'Nantianhuang'.

It is worth pointing out that the blight-resistant banana varieties that have been selected and bred need to be tested for adaptation to different sites of origins in order to be applied in the production practice. By such an application and promotion of the selected disease-resistant varieties, 'Nantianhuang' has been already planted in Yunnan, Guangdong and Hainan, 'Zhongjiao' has been cropped in Guangdong, and 'Guijiao' has been planted in Guangxi and other places in a large area. However, compared with the widely grown but Panama disease-susceptible 'Cavendish Brazilian', these resistant varieties tend to show varied characteristics and have long reproductive period and different fertilizer and water requirements. Therefore, it is necessary to explore the cultivation and management measures for these selected resistant varieties and implement the principle that "excellent varieties must be accompanied by corresponding excellent cultivation methods", which could produce the banana products accepted by the market. Given that different disease-resistant varieties exhibited distinct advantages in controlling Panama disease in different regions, it remains necessary to carry out rigorous selection and breeding of disease-resistant varieties, which could show advantages of high and stable resistance and desirable traits with high yield and quality across all of the banana production area in China.

4.2 Chemical Control Methods

Chemical control is an important method of plant disease control. Many scientists all over the world have carried out toxicity determination *in vitro* and field efficacy tests of chemical agents against the pathogen (Foc4). Yang et al. (2019) determined the virulence of seven soil disinfectants (dazomet, 1, 3-dichloropropene, allyl isothiocyanate, propylene oxide, glutaraldehyde, trichloroisocyanuric acid, and chlorobromoisocyanuric acid) against Foc4 *in vitro*. All seven soil disinfectants exhibited inhibitory effects on the growth of Foc4 mycelium to a certain extent. Duan et al.

(2015) reported that when the mass concentration of chlorine dioxide was larger than 400 mg/L, the Foc4 could be completely killed after 24 h of treatment even when the pathogen density reached at 10^5 CFU/cm^3 in soil. Furthermore, other fumigations, *i.e.*, calcium carbamide, imazethapyr, carbendazim, fernasan, and dazomet, have been shown to be effective in inhibiting the growth of Foc4 mycelium in soil (Huang et al. 2017). Interestingly, several green fumigations such as liquid ammonia, ammonia water, ammonium bicarbonate, were also found to be effective in inhibiting Foc4 in soil (Shen et al. 2018, 2019a, b). Although a large number of effective pesticides have been reported for controlling banana Panama disease control have been reported, further research is needed to screen pesticides that are easily absorbed to banana plants, determine optimal application methods, and define the rational timing. This is particularly important because Foc4 could survive in soil via different forms across diverse geographical sites, highlighting the need to detangle the general microbial features responding to disease suppression.

4.3 Biological Control as a Management Strategy

Screening of antagonistic microorganisms and investigating their inhibitory effect and mechanism on Foc4 are hotspots of current research on banana Panama disease control. Many antagonistic strains with excellent ability to inhibit the pathogen have been screened and these results are available in China. For example, Zhou et al. (2017) successfully isolated 93 bacterial strains and 133 strains of actinomycetes antagonistic to Foc4 from banana cropped soil. Tian et al. (2021) isolated a strain identified as *Bacillus amyloliquefaciens* GKT04 from the banana roots of 'Guijiao No. 9', which could reduce the disease incidence of banana Panama disease on banana seedlings by 49% under greenhouse condition. Wang et al. (2016) have successfully screened a strain of *B. amyloliquefaciens* W19 antagonistic to Foc4 from banana mono-cropped soils by producing secondary metabolites, such as *iturin* and *bacillomycin* D.

Recently, fungi have been identified as alternative biocontrol agents against banana Panama disease, even showing more inhibitory effects on Foc4 compared to bacterial biocontrol strains. For example, Ye et al. (2019) found that a biocontrol product synthetized by three *Trichoderma* strains (Tr309T, Tr121, Tr14) exhibited up to 87% disease control under both greenhouse and field conditions. Zhu et al. (2021) has successfully isolated a thermotolerant *Trichoderma* strain JS84, classified as *Trichoderma citrinoviride*, which could significantly suppress banana Panama disease incidence, improve the quality of banana fruit and also, soil nutrient condition. Here are two case studies demonstrating the effects and possible mechanism of a biocontrol strategy to mitigate banana Panama disease.

4.3.1 Leveraging Beneficial Microbes to Combat Banana Panama Disease

It is well known that the dominant soil microbial communities responsible for disease suppression often differ across different crop types and soil properties (Raaijmakers and Mazzola 2016; Kwak et al. 2018). Therefore, there is an urgent need for comprehensive research on the microbiome of disease-suppressive soils.

Lv et al. (2024) conducted a field survey in 2019 to explore the soils potential suppressive to banana Panama disease among the banana main producing area. By combining real-time quantitative PCR with plant-based pathogen inoculation assays, multiple co-located orchards were surveyed, leading to the identification and validation of two potentially suppressive soils, sampled from JF (Jianfeng town) and LG (Lingao county). Microbiome analyses further revealed key microbial taxa, *e.g.*, *Massilia*, *Brevundimonas*, and *Pseudomonas*, may play key role in suppressing *Fusarium* pathogen. Subsequently, culturable isolates belonging to *Brevundimonas* and *Pseudomonas* were obtained, and their functional characteristics were investigated through microcosm and greenhouse experiments. These isolates demonstrated strong antagonistic activity against pathogens and were able to stimulate plant immune responses. A synthetic microbial consortium composed of the culturable *Brevundimonas* S6 and *Pseudomonas* S30 significantly enhanced the suppression ability against soilborne pathogens. This effect was associated with increased activity of plant disease-resistance enzymes, upregulation of NRPS-related genes, and modulation of the rhizosphere microbial community.

Taken together, these results indicate that disease-suppressive soils are enriched with beneficial microorganisms, and the inoculation of beneficial microorganisms could be crucial for establishing a beneficial plant-microbe holobiont. This, in turn, helps inhibit pathogen colonization, supports the normal functions of the microbial community, and maintains plant health. However, it is important to note that to fully understand the interactions between beneficial microbial groups (*e.g.*, *Brevundimonas* and *Pseudomonas*), more following studies are needed to explore the microbial synergies within the context of the rhizosphere.

4.3.2 Biofertilizer Containing *Bacillus* sp. Application Enhances Plant Disease Suppression

The biofertilizer application has been shown to be effective in enhancing plant pathogen resistance bases on lots of previous reports. However, it remains unclear whether such improvements are caused by the specific practice of introduced beneficial inoculants, indigenous beneficial consortia in organic fertilizer or the physicochemical characteristics of the compost matrix. To elucidate the operational mechanisms of biofertilizers, we conducted a longitudinal study monitoring Panama disease in banana plantations and concomitant soil microbiome shifts across three consecutive cropping seasons. Four experimental treatments were established: (1) biofertilizer containing *Bacillus amyloliquefaciens* W19, (2) organic fertilizer, (3) γ-irradiated organic fertilizer, and (4) γ-irradiated organic fertilizer re-inoculated with *B. amyloliquefaciens* W19 (Tao et al. 2020).

Sole application of organic amendments demonstrates limited efficacy in pathogen control, with documented cases of phytopathogen proliferation enhancement. *Bacillus amyloliquefaciens* W19 emerged as the determining biocontrol agent for effective suppression of banana Panama disease. As summarized in the conceptual model describing the mode of action of biofertilizer application in suppressing *Fusarium* wilt, the biological agent *B. amyloliquefaciens* W19 is capable of colonizing the rhizosphere and exerting its effects through the following mechanisms:

(A) Direct inhibition of pathogen growth via production of antimicrobial compounds;
(B) Induction of microbiome alterations (particularly shifts in bacterial composition) leading to reduced pathogen density;
(C) Synergistic activation of specific beneficial plant-associated microorganisms within the rhizosphere microbiome;
(D) Potential promotion of multi-species biofilm formation with specific plant-beneficial bacterial genera (*e.g.*, *Pseudomonas*);
(E) Limitation of fungal pathogen infectivity through direct or indirect pathways that restrict root colonization capacity.

The experiments we designed to enable systematic dissection of the differential contributions of distinct components in the bio-organic fertilizer to disease suppression efficacy.

The observed disease suppression in these treatments was mediated through resident soil microbiome restructuring, particularly via selective enrichment of specific *Pseudomonas* spp. We systematically investigated this correlation between Bacillus amendment and indigenous *Pseudomonas* spp. through controlled laboratory assays and replicated pot trials, which revealed synergistic biofilm assembly between inoculated *Bacillus* and indigenous Pseudomonas strains that likely acted as a plant-beneficial consortium to enhance plant-protective capacity against pathogen invasion. Together, our integrated analysis demonstrates that bioorganic fertilizer efficacy stems from dual mechanisms: (a) direct introduction of biocontrol agents, and (b) microbiome-mediated amplification of suppressive functions (Fig. 4.2). We have demonstrated that the action of bioorganic fertilizer is a product of the biocontrol inoculum within the organic amendment and its impact on the resident soil microbiome (Fig. 4.2). These findings establish a mechanistic framework for the design of more efficient biofertilizers that optimize microbiome-mediated soil functionality.

4.3.3 Biofertilizer Containing *Trichoderma* sp. Application Drives Biocontrol of Banana Panama Disease

Host-associated fungi contribute to plant defense against pathogens, with solid evidence demonstrating that these microbial communities can be manipulated through probiotic introduction to enhance disease-suppressive effects. However, a pervasive

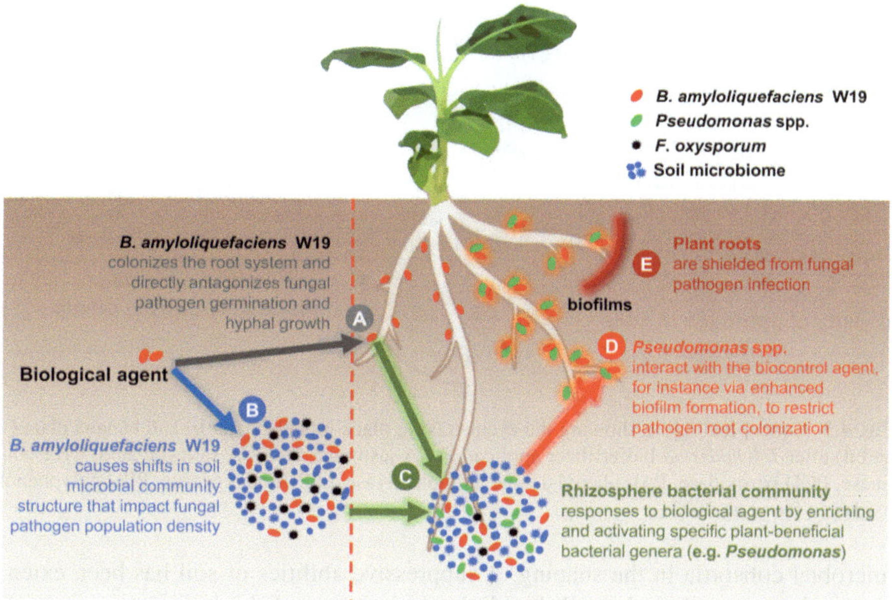

Fig. 4.2 A conceptual figure showing the changes in microbial community composition and activities that restrict fungal pathogen growth and subsequent plant infection induced by biofertilizer application (Tao et al. 2020)

knowledge gap persists regarding the mechanisms governing probiotic efficacy, which critically impedes the development of robust disease mitigation strategies.

Tao et al. (2023) conducted a tri-seasonal microcosm experiment to evaluate the effects of *Trichoderma*-enriched biofertilizer application in soils susceptible to *Fusarium* wilt of banana. The non-treated control group received no microbial inoculation. The study systematically monitored fungal community dynamics through integrated high-throughput sequencing and culture-based approaches, while investigating their potential associations with plant health. The experiments revealed that biofertilizer amended with *Trichoderma guizhouense* NJAU4742, a strain with demonstrated biocontrol efficacy against wilt disease, achieved a 72% reduction in disease incidence alongside significantly increased banana biomass. This suppression was mainly attributed to fungal microbiome restructuring, including decreased density of pathogen and enrichment of pathogen antagonistic fungi (*e.g.*, *Humicola* spp.). These shifts coincided with expanded microbial carbon resource utilization potential, features that synergistically contribute to pathogen suppression. Tao et al. (2023) further demonstrated the disease-suppressive function of the *Trichoderma*-*Humicola* consortium, with results suggesting that niche overlap with pathogens and induced plant systemic resistance may mechanistically drive the observed biocontrol outcomes.

Integrated analyses demonstrated that fungal inoculants restructure the resident soil fungal microbiome composition and functionality, establishing disease-suppressive states against soil-borne pathogens (Fig. 4.3). While the role of

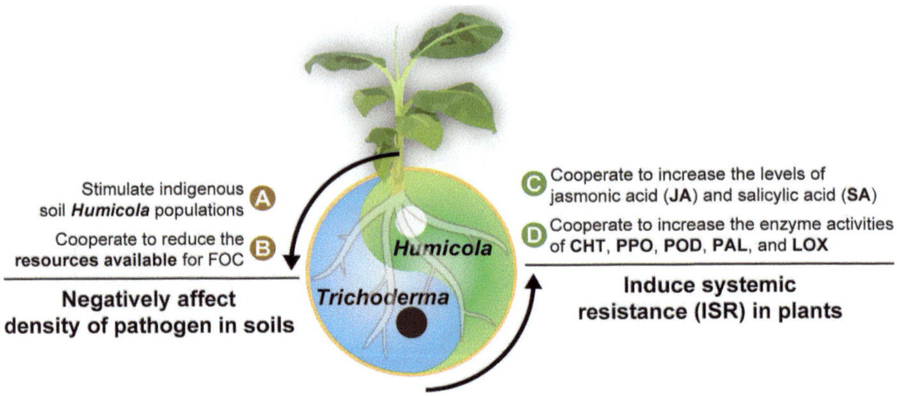

Fig. 4.3 Conceptual figure showing the events taking place in soil (**a** and **b**) and banana plant (**c** and **d**) after *Trichoderma* biofertilizer application. JA jasmonic acid, SA salicylic acid, CHT chitinase, POD peroxidase, PAL phenylalanine ammonia lyase, LOX lipoxygenase, PPO polyphenol oxidase. (Tao et al. 2023)

microbial consortia in the shaping of suppressive abilities of soil has been extensively documented and well highlighted, impacts of biofertilizer containing *Trichoderma* sp. application on fungal microbiome and specific interactions within the fungal community have emerged as important drivers of plant health, which provide novel insights toward the development of effective and sustainable solutions to achieve crop protection.

It is worth mentioning that biocontrol agents and their derived products (*i.e.*, biofertilizer) have displayed excellent effects in controlling banana Panama disease. However, several issues should be addressed to enhance the stability of biological control agent application in the field, including the production of biofertilizers containing synthetized antagonists, extending the survival and activity of antagonists in the products, and the appropriate methods of application into banana soil.

4.4 Agricultural Practices

The most usually adopted agricultural management strategies such as soil remediation, crop rotation, application of organic fertilizer (bio-fertilizer) and/or new-type of fertilizer, intercropping, have been reported to be effective in controlling banana Panama disease. For example, Zeng et al. (2019) reported that sugarcane rotation reduced the incidence of banana Panama disease in continuously cropped banana field from 49.15% to 17.86% in the first year, and further reduced to 1.79% after the two- and three-year rotations. Furthermore, rotation of Chinese leek, chili peppers, wax gourd, rice, cassava, maize, pineapple, ginger and cucumber, have also been found to be effective in controlling this disease. As for intercropping, sweet potato, peanut, soybean, green manure (*i.e.*, *Trifolium repens*, mung bean), and maize have been reported as useful crops for intercropping in reducing banana Panama disease in China. Reductive soil disinfestation (RSD) together with organic material application has been proposed as an alternative method to control banana Panama disease by Huang et al. (2015). Moreover, several soil additives, such as alkaline chemical fertilizer, lime, biochar, were also found to be helpful in reducing this disease incidence.

It is worth noting that the field effect of agricultural measures is not stable and can only alleviate or reduce the occurrence of banana wilt to a certain extent, and the effect of any single agricultural control measure is not satisfactory. However, several reports have shown the advantages of controlling this disease. Here is an introduction related to the effects and possible mechanisms of several promising agricultural practices in mitigating banana Panama disease in China.

4.4.1 Soil Fumigation Coupled with Biofertilizer Application Suppressed Banana Panama Disease via Microbiome Manipulation

Ammonia fumigation serves as a green soil treatment that controls Panama disease (*Fusarium* wilt of banana) by either disrupting fungal mycelia or inhibiting conidial germination of Foc4. However, the holistic impacts of ammonia fumigation on soil bacterial and fungal community composition, as well as the post-application shifts following biofertilizer amendment, remained poorly understood. To address this knowledge gap, Shen et al. (2019a, b) conducted a two-season greenhouse experiment designed to investigate microbial community divergence induced by soil fumigation and biofertilizer application, while elucidating underlying disease-suppressive mechanisms derived from these community shifts.

According to greenhouse experiments conducted by Shen et al. (2019a, b), the application of biofertilizer after ammonia fumigation significantly reduced the incidence of banana Panama disease by approximately 55% compared to the control fertilizer (cattle manure) applied in non-fumigated soil, while also promoting plant biomass growth. Results from organic amendments applied during the fumigation phase and subsequent banana replanting stage demonstrated that in previously heavily infested soils, post-fumigation application of biofertilizer effectively controlled banana Panama disease and enhanced banana growth under greenhouse conditions. The observed disease suppression may be attributed to two mechanisms: in the first, the direct inhibition of pathogen populations through the combined effects of fumigation and fertilization, and in the second, comprehensive inhibitory effects mediated by soil microbiome regulation by soil abiotic factors (*e.g.*, increasing pH and organic matter). More specifically, the application of biofertilizer following ammonia fumigation modulated the soil microbiome in continuous banana cropping fields by: (1) altering soil bacterial community structure, and (2) reducing total fungal population and the relative abundance of pathogen, while simultaneously activating native beneficial microbial consortia (including Actinobacteria phylum members like *Actinoplanes*, Firmicutes phylum members such as *Paenibacillus*, along with *Bradyrhizobium*, *Nitrosomonas*, and *Nitrobacter*).

4.4.2 Soil Pre-fumigation Improved the Disease Suppressiveness of Biofertilizer to Banana Fusarium Wilt Disease by Re-shaping Soil Microbiome

An integrated approach combining soil fumigation with subsequent biofertilizer application has been demonstrated as effective in controlling banana Panama disease. Shen et al. (2018) conducted a two-seasonal pot experiment to investigate the suppressive efficacy of biofertilizer following lime-ammonium bicarbonate (LA) fumigation against banana Panama disease, along with its impacts on soil microbial community composition.

According to findings investigated by Shen et al. (2018), the application of biofertilizer after LA fumigation significantly reduced the incidence of Panama disease by 80% compared to the untreated control group that biofertilizer was directly applied to non-fumigated soil. Non-metric multidimensional scaling (NMDS) and Venn diagrams analysis revealed that biofertilizer application after fumigation significantly altered soil microbial community composition. In detail, LA fumigation markedly decreased the abundance of pathogen (Foc4), and subsequent biofertilizer application further amplified this suppression. Additionally, soil native microorganisms (e.g., *Bacillus*, *Pseudomonas*, and *Mortierella*) were found to be associated with disease suppression. Redundancy analysis (RDA) and variation partitioning analysis demonstrated that biofertilizer application after fumigation significantly increased the value of soil pH, contents of total organic carbon, available phosphorus and available potassium levels, which may represent critical factors in reshaping soil microbial communities. The observed disease suppression may be attributed to a synergistic inhibitory effect arising from fumigation and fertilization through the improvement of soil properties (*e.g.*, pH and total organic carbon).

More specifically, biofertilizer application after soil fumigation manipulated the microbiome in continuous banana cropping soils through: A) Altering soil bacterial community composition; B) Enhancing phylogenetic diversity of both bacteria and fungi; C) Reducing the relative abundance of Foc4 pathogen; D) Activating native beneficial microbial consortia (*e.g.*, *Bacillus*, *Pseudomonas*, *Nocardioides*, and *Mortierella*). Nevertheless, functional parameters for predicting disease suppression efficacy following biofertilizer application remain to be clearly defined, and this will be a key focus of our future research.

4.4.3 Soil Fumigation Coupled with Biofertilizer Application Steered Banana Rhizosphere to Assemble a Unique Microbiome Against Panama Disease

A previous study (Shen et al. 2018) showed that microbial community of bulk soil could be manipulated by lime and ammonium bicarbonate fumigation followed by biofertilizer application. However, the assembly mechanisms of microbial communities in the rhizosphere soil that suppress banana *Fusarium* wilt disease remain unclear. To address this, Shen et al. (2019a, b) conducted a two-seasonal pot experiment by using high-throughput sequencing and quantitative PCR techniques to investigate the assembly patterns of rhizosphere microbiomes associated with disease suppression. It was found that compared to biofertilizers applied in non-fumigated soils, biofertilizer applied to soils fumigated with lime and ammonium bicarbonate significantly reduced the abundance of Foc4 pathogen in the banana rhizosphere. Further principal coordinate analysis (PCoA) revealed that biofertilizers applied in fumigated soils re-constructed the rhizobacterial community composition by enhancing phylogenetic relatedness and inducing native microorganisms such as *Pseudomonas*, *Lysobacter*, and *Bacillus*. Co-occurrence network analysis demonstrated that potential taxa involved in disease suppression exhibited more tightly interconnected relationships in disease-suppressive soils.

In summary, the application of biofertilizer following lime and ammonium bicarbonate fumigation triggered the assembly of a beneficial-microbe dominated community in the banana rhizosphere, thereby suppressing banana *Fusarium* wilt disease. The observed disease-suppressive effects may be attributed to the combined suppression effects generated by fertilization and fumigation probably through the following mechanisms: (A) altering soil community structure; (B) enhancing phylogenetic relatedness; (C) reducing the relative abundance of Foc4 pathogen; (D) stimulating indigenous beneficial microbial communities (*e.g.*, *Pseudomonas*, *Lysobacter*, and *Bacillus*); and (E) fostering interconnections and enhanced organization among potential taxa involved in disease suppression.

4.4.4 Promoting Soil Suppressive Ability Against Banana Panama Disease via Enriching the Specific Fungal by Crop Rotation

Crop rotation could significantly affect the soil health and thereby the sustainable crop production. Pineapple cultivation in the high incidence soil was observed that it can significantly decrease the pathogen (Foc4) density and disease incidence in Hainan, China (Wang et al. 2015). Considering that the mechanisms of pineapple rotation in sustaining disease suppression is still unclear, a banana-pineapple rotation system has been investigated by Yuan et al. (2021) as a representative rotation case to introduce the potential ability of a microbial-mediated disease-suppression in the Foc4 highly infested soil.

The study (Yuan et al. 2021) conducted by both pot and field experiment demonstrated that pineapple-banana rotation significantly reduced Foc4 population and Panama disease incidence. The suppression to Foc4 was associated with detectable shifts in soil-resident fungal taxa during the pineapple cultivation phase. Furthermore, the abundance of taxa affiliated with *Talaromyces pinophilus* and *Clonostachys rossmaniae* markedly increased in the soil during the pineapple season. Moreover, *T. pinophilus* was also enriched in the banana rhizosphere during the subsequent banana season. Subsequently, researchers validated the ability of *T. pinophilus* and *C. rossmaniae* to reduce the Foc4 pathogen in banana rhizosphere soil through fungal isolation and pot inoculation assays, confirming their role in suppressing Panama disease.

In summary, the study (Yuan et al. 2021) provides empirical evidence for how crop rotation shapes soil-resident microbiomes and enhance disease-suppressive capacity (Fig. 4.4). Moreover, the research underscores the importance of understanding crop rotation-mediated soil biological dynamics and validates potential mechanisms of disease suppression through the targeted steering of protective soil microbiomes in agroecosystems. Therefore, exploring the underlying mechanisms that mediate the assembly of beneficial microbiomes and elucidating the biological interactions among microbial taxa involved in disease suppression are critical for advancing our ability to engineer microbiomes in field settings.

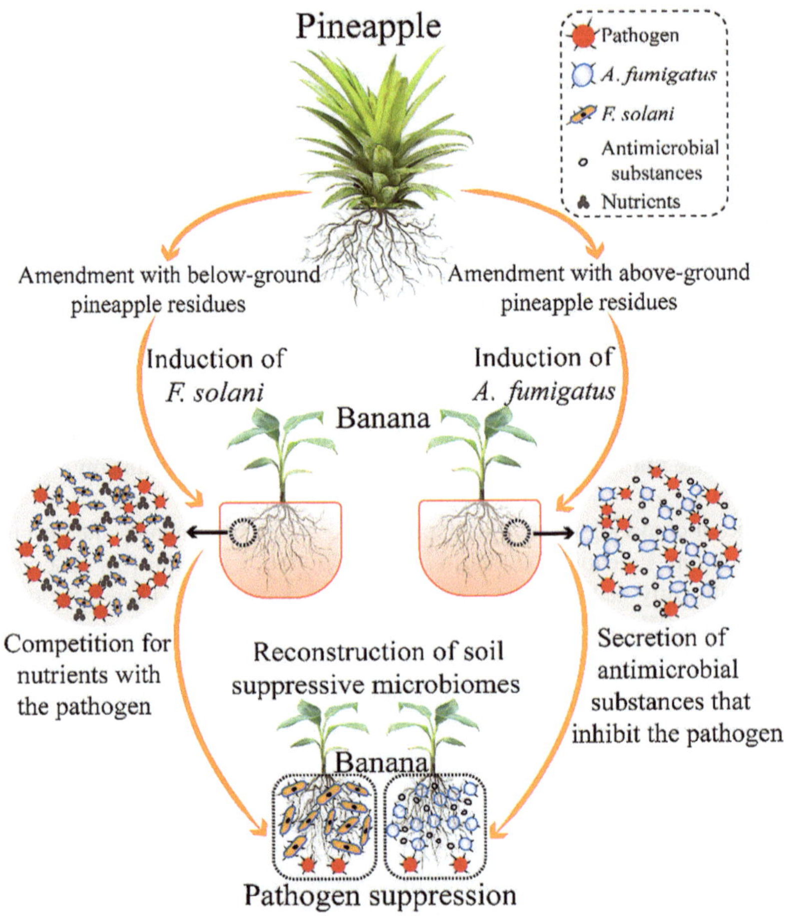

Fig. 4.4 The conceptual model instructing the function of suppressive taxa influencing the pathogen and the *Fusarium* wilt disease. (Yuan et al. 2021)

4.4.5 Enhancing Soil Suppressiveness Against Banana Panama Disease through Manipulating Fungal Communities via Plant Residues Amending

Mining soils with disease-suppressive functions represents a promising approach to protect plants from soil-borne diseases. The application of crop rotation and incorporating plant residues into soil are known to alleviate pathogen pressure by altering the dynamics of soil biological and physicochemical properties. However, research remains limited on the extent to which specific soil amendment practices involving plant residues promote the formation of beneficial microbiomes. Here, how introducing pineapple residues into soils severely affected by banana *Fusarium* wilt reducing pathogen pressure through microbiome improvement was explored by Yuan et al. (2021).

In soils with high disease incidence, above- and below-ground pineapple residues amendment significantly reduced pathogen populations and consequently lowered disease occurrence. The formation of disease-suppressive soils was primarily associated with shifts in specific fungal taxa such as *Aspergillus fumigatus* and *Fusarium solani*, both of which exhibited marked suppressing effects against the Foc4 pathogen. The antagonistic roles of these beneficial microbes were further validated through cultivation experiments, revealing their suppression mechanisms via direct secretion of antimicrobial substances and indirect competition. The disease-suppressive potential of these fungal strains was confirmed through microbial inoculation in pot experiment.

These results comprehensively elucidate how the addition of plant residues to soil induces shifts in the soil microbiome, thereby directly influencing disease suppression processes. After pineapple residues are added to the soil, they can reduce the abundance of pathogenic by increasing the number of certain antagonistic fungi, thereby having a negative impact on the number of pathogenic and the incidence of *Fusarium* wilt. In summary, this study (Yuan et al. 2021) provides a successful paradigm demonstrating how targeted agricultural management strategies can improve soil microbiomes, stimulate disease-suppressive soils, and ultimately suppress soil-borne diseases (Fig. 4.5). Furthermore, the research offers novel pathways for exploring agricultural practices that sustainably and directly enhance beneficial outcomes such as soil health and crop productivity.

4.4.6 Stimulating Soil Suppressiveness Against Banana Panama Disease Through Manipulating Soil Microbiome by Green Manure Intercropping

Although intercropping was well recognized as an effective strategy in controlling banana Panama disease, the mechanism of intercropping in sustaining disease suppression still remains unclear. The field and pot experiments were conducted by Yang et al. (2022) to evaluate the effects of five intercropped green manure crops

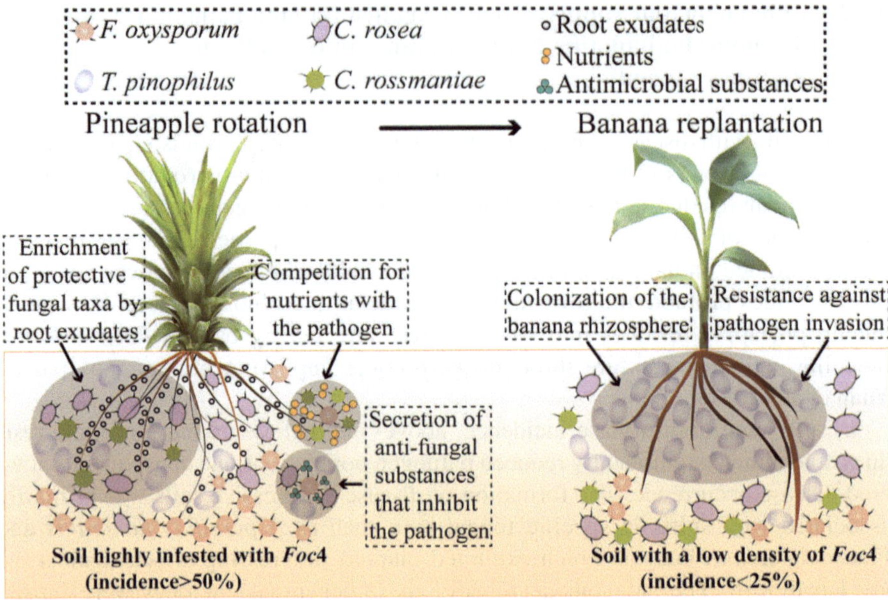

Fig. 4.5 A concept map of the potential mechanism by which crop rotation promotes the emergence of inhibitory soil by enhancing specific fungi capable of combating pathogens. DI disease incidence. (Yuan et al. 2021)

(*Crotalaria acicularis*, *Melilotus officinali*, *Sesbania sesban*, *Trifolium repens* and *Vicia villosa*) on banana *Fusarium* wilt through field and pot experiment.

The study by Yang et al. (2022) demonstrated that intercropping banana with five green manure crops all significantly reduced the incidence of soil-borne *Fusarium* wilt while decreasing fungal richness and diversity in the soil. Among the intercropping treatments, *V. villosa* (hairy vetch) and *T. repens* (white clover) exhibited superior disease-suppressive effects compared to other treatments. All five intercropping systems showed increased abundance of *Aspergillus* and reduced populations of culturable Foc4 pathogen relative to banana monoculture. Structural equation modeling (SEM) revealed that changes in soil organic matter and NH_4^+-N played pivotal roles in both the reduction of *Fusarium* wilt incidence and shifts in soil microbial communities. In summary, this research provides insights into green manure intercropping as a strategy to suppress soil-borne *Fusarium* wilt and elucidates its microbial mechanisms within the context of agricultural soil management. Future research may focus on elucidating soil microbe-plant interactions under intercropping conditions to better understand the mechanisms supporting the auxiliary control of banana Panama disease.

4.4.7 Resistant Varieties Shape Rhizosphere Fungal Communities to Suppress Pathogens

Despite extensive efforts by scientists to control banana Panama disease, breeding resistant banana varieties remains the most fundamental safeguard in managing this disease. Significant research has been devoted to developing resistant banana varieties, leading to a diverse range of options that exhibit varying levels of resistance, from highly susceptible to highly resistant as reported by Liu et al. (2023). Quantitative analysis of rhizosphere soil pathogen of these different banana varieties revealed that a negative correlation between pathogen abundance and the level of host resistance, with more resistant varieties exhibiting lower relative abundance of the pathogen. Previous studies on the mechanisms of disease resistance have primarily focused on the plant immune system and resistance genes. However, the widespread adoption of resistant varieties has been hampered by challenges such as regional adaptability and the erosion of resistance due to pathogen evolution (McDonald and Linde 2002). As the plant's "second genome", the rhizosphere microbiome confers fitness advantages to the plant performance and health (Trivedi et al. 2020). In response to soil-borne pathogen threats, plants are capable of recruiting specific microbial populations and reshaping the composition and functionality of their rhizosphere microbiome to support plant health (Yin et al. 2020). Therefore, enhancing soil disease suppression and promoting positive interactions between the rhizosphere and banana plants are essential prerequisites for the sustainable development of the banana industry.

Liu et al. (2023) further revealed that the composition of rhizosphere fungal communities varied significantly among banana varieties with differing levels of resistance to disease, highlighting the strong association between rhizosphere microbiomes and plant disease resistance. A more detailed analysis of the rhizosphere microbial community composition, combined with cultivation-dependent approaches, revealed that disease-resistant banana varieties significantly enriched *Trichoderma* spp. and *Penicillium* spp. in their rhizosphere, which greatly reduced pathogen colonization and contributed to maintaining plant health. Importantly, the recruitment of these beneficial fungi was driven by specific differential rhizosphere metabolites, such as shikimic acid and D-(−)-ribofuranose. Taken together, the observed disease resistance in plants is largely attributed to their ability to selectively recruit disease-suppressive fungal communities in the rhizosphere (Fig. 4.6). The recruitment traits of disease-resistant varieties provide valuable insights for the development of disease-suppressive soils.

5 Summary and Outlook

Banana is an important cash crop in South China, leading to the increase of employment and income of personnel in related industries. However, in the last two decades, banana Panama disease has become more and more serious due to changes in

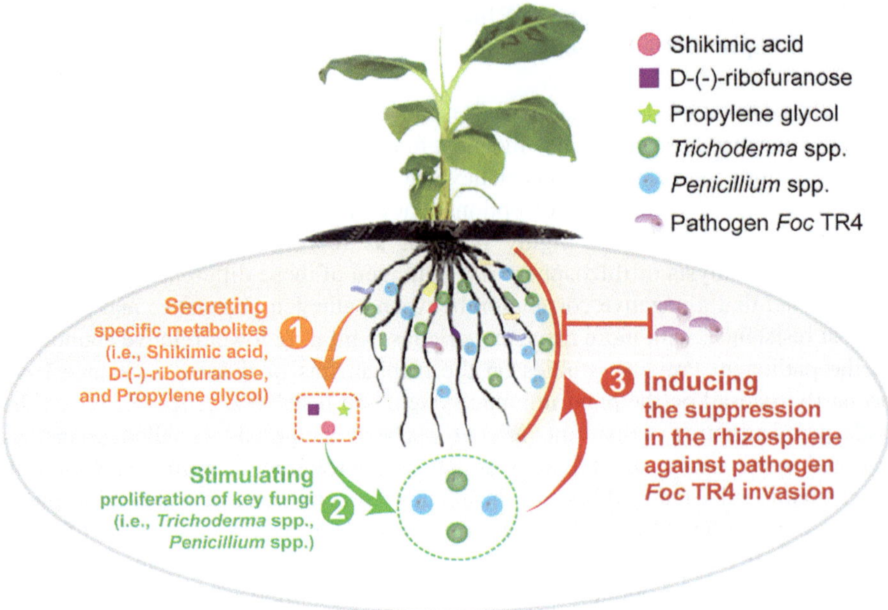

Fig. 4.6 Schematic representation of rhizosphere metabolite-mediated interactions among plants, pathogens, and beneficial fungal taxa. (Liu et al. 2023)

farming practices and production technologies in China, causing huge losses to the banana industry. Although a range of management strategies has proven to be effective in controlling banana Panama disease, inducing the soil suppressiveness against banana Panama disease has currently become the most promising and long-lasting method in overcoming this disease. Knowing that soil is a reservoir of soil-borne pathogens, and rhizosphere microbial community determines the soil and crop health, manipulating the soil and rhizosphere microbiome to enhance soil suppression is a feasible means of preventing and controlling banana Panama diseases.

The current research on triggering the soil microbiome to control banana Panama disease has shown excellent efficiency in China, and a number of biocontrol microbial strains have been successfully screened. Furthermore, the related mechanism of action has also been successfully revealed. However, some issues, such as the short shelf-life of biocontrol products, and unsatisfactory preventive effect in the field, still need to be addressed in the future. Therefore, the breeding of excellent strains and stability of the derived products with longer shelf-life are the key issues that will determine the research and development of superior biocontrol agents should receive more attention in the future. Moreover, it is necessary to push the limits of screening and isolating biocontrol resources to search for efficient and environment-friendly prevention and control methods that should be studied through in-depth research.

The following aspects may be helpful in exploring the extended shelf life for effective ways in mitigating banana Panama disease:

1. **Disentangling the importance of soil biotics in maintaining soil health**. The roles of soil microbiome, especially bacteria, in banana soils with excellent suppressiveness ability against *Fusarium* wilt disease in China has been well deciphered during the recent years. However, with the recognition of the value of soil protozoa and nematodes in determining soil and crop health, role of soil microbial food web in sustaining disease suppression should be paid more attention in the future.
2. **Comprehensive and thorough investigation of the interaction mechanism between banana Panama disease and rhizosphere microbiome**. The differences in root exudates, rhizosphere and root microorganisms, and soil physicochemical properties between healthy and diseased banana plants should be further clarified. The effects of environmental factors on the occurrence and development of banana Panama disease by integrating high-throughput sequencing and multi-omics (metagenomics, metatransgenomics and metabolomics) should be well investigated to provide theoretical bases for the prevention and control of the disease.
3. **Developing highly efficient and stable biocontrol agents with wider application**. Exploring the appropriate temperature, pH, carbon and nitrogen sources, inorganic salts, time and concentration required for the cultivation of biocontrol strains and preparation of biocontrol preparations, are all required and will be helpful to establish a scientifically standardized and unified method of screening and isolating biocontrol agents and preparation of biocontrol formulations, which finally could enhance the stability of the efficacy of the preparations in the field. In addition, the construction of banana endophytes carrying disease resistance genes and their colonization and expression after inoculation into bananas for enhancing banana's resistance to wilt is also a direction for the use of endophytes as biocontrol strains in banana wilt control and application.
4. **Rhizosphere immunity is proposed as a promising concept for supporting plant health**. Rhizosphere immunity, as a holistic conceptual framework, is based on ecosystem theory and integrates soilborne disease management theories and technological measures centered on plants, soils, microbiomes, and pathogens. It emphasizes the importance of not only understanding each component of inter-root immunity independently but also paring to show how the components interact with and influence each other, thus, objectively assessing their contributions to inter-root immunity and plant health. To achieve these goals, it is necessary to divide the work among those working in different disciplinary areas and promote multidisciplinary collaboration to identify scientific issues related to inter-root immunization and to promote the cross-fertilization and application of knowledge and concepts from different disciplines.

References

Berendsen RL, Pieterse CMJ, Bakker PAHM (2012) The rhizosphere microbiome and plant health. Trends Plant Sci 17:478–486

Butler D (2013) Fungus threatens top banana. Nature 504:195–196

Cao Y, Ou Y, Lv N, Tao C, Liu H, Li R, Shen Z, Shen Q (2024) Soil nutrient levels are associated with suppression of banana *Fusarium* wilt disease. Soil Ecology Lett 6:240247

Cook RJ, Rovira AD (1976) The role of bacteria in the biological control of *Gaeumannomyces graminis* by suppressive soils. Soil Biol Biochem 8:269

Dita MA, Garming H, van den Bergh I, Staver C, Lescot T (2011) Banana in Latin America and the Caribbean: current state, challenges and perspectives. In: International ISHS-ProMusa symposium on bananas and plantains – towards sustainable global production and improved use, Salvador, Brazil

Duan Y, Pang Z, Chen J, Hu Y, Hu H, Xie J (2015) A preliminary study on the effectiveness of chlorine dioxide in the control of banana wilt in soil, South China. Fruits 44:74–77. +81. (in Chinese)

Henry AW (1931) The natural microflora of the soil in relation to the foot-rot problem of wheat. Can J Res 4:69–77

Heslop-Harrison JS, Schwarzacher T (2007) Domestication, genomics and the future for Banana. Ann Bot 100:1073–1084

Huang XQ, Liu LL, Wen T, Zhu R, Zhang JB, Cai ZC (2015) Illumina MiSeq investigations on the changes of microbial community in the *Fusarium oxysporum* f.sp *cubense* infected soil during and after reductive soil disinfestation. Microbiol Res 181:33–42

Huang ZF, Liuyang RQ, Dong CY, Lei YF, Zhang AQ, Lin YL (2017) Polymeric quaternary ammonium salt activity against *Fusarium oxysporum* f. sp cubense race 4: synthesis, structure-activity relationship and mode of action. React Funct Polym 114:13–22

Janvier C, Villeneuve F, Alabouvette C, Edel-Hermann V, Mateille T, Steinberg C (2007) Soil health through soil disease suppression: which strategy from descriptors to indicators? Soil Biol Biochem 39:1–23

Jeger MJ, Eden-Green S, Thresh JM, Johanson A, Waller JM, Brown AE (1995) Banana diseases. In: Bananas and plantains. Chapman & Hall, London

Kwak M-J, Kong HG, Choi K, Kwon S-K, Song JY, Lee J, Lee PA, Choi SY, Seo M, Lee HJ, Jung EJ, Park H, Roy N, Kim H, Lee MM, Rubin EM, Lee S-W, Kim JF (2018) Rhizosphere microbiome structure alters to enable wilt resistance in tomato. Nat Biotechnol 36:1100–1109

Liu S, Tao C, Zhang L, Wang Z, Xiong W, Xiang D, Sheng O, Wang J, Li R, Shen Z, Li C, Shen Q, Kowalchuk GA (2023) Plant pathogen resistance is mediated by recruitment of specific rhizosphere fungi. ISME J 17:931–942

Lv N, Ravanbakhsh M, Ling S, Ou Y, Tao C, Liu H, Li R, Shen Z, Shen Q (2024) Harnessing key bacteria from suppressive soil to mitigate banana Panama disease. Biol Fertil Soils 60:845–862

McDonald BA, Linde C (2002) Pathogen population genetics, evolutionary potential, and durable resistance. Annu Rev Phytopathol 40:349–379

Meng QX, Yin JF, Rosenzweig N, Douches D, Hao JJJ (2012) Culture-based assessment of microbial communities in soil suppressive to potato common scab. Plant Dis 96:712–717

Peng HX, Sivasithamparam K, Turner DW (1999) Chlamydospore germination and Fusarium wilt of banana plantlets in suppressive and conducive soils are affected by physical and chemical factors. Soil Biol Biochem 31:1363–1374

Ploetz RC (2006) Fusarium wilt of banana is caused by several pathogens referred to as *Fusarium oxysporum* f. sp *cubense*. Phytopathology 96:653–656

Raaijmakers JM, Mazzola M (2016) *Soil immune responses* soil microbiomes may be harnessed for plant health. Science 352:1392–1393

Shen Z, Ruan Y, Xue C, Zhong S, Li R, Shen Q (2015) Soils naturally suppressive to banana Fusarium wilt disease harbor unique bacterial communities. Plant Soil 393:21–33

Shen Z, Xue C, Taylor PWJ, Ou Y, Wang B, Zhao Y, Ruan Y, Li R, Shen Q (2018) Soil pre-fumigation could effectively improve the disease suppressiveness of biofertilizer to banana Fusarium wilt disease by reshaping the soil microbiome. Biol Fertil Soils 54:793–806

Shen Z, Xue C, Penton CR, Thomashow LS, Zhang N, Wang B, Ruan Y, Li R, Shen Q (2019a) Suppression of banana Panama disease induced by soil microbiome reconstruction through an integrated agricultural strategy. Soil Biol Biochem 128:164–174

Shen ZZ, Wang BB, Zhu JX, Hu HW, Tao CY, Ou YN, Deng XH, Ling N, Li R, Shen QR (2019b) Lime and ammonium carbonate fumigation coupled with bio-organic fertilizer application steered banana rhizosphere to assemble a unique microbiome against Panama disease. Microb Biotechnol 12:515–527

Shen Z, Thomashow LS, Ou Y, Tao C, Wang J, Xiong W, Liu H, Li R, Shen Q, Kowalchuk GA (2022) Shared core microbiome and functionality of key taxa suppressive to banana fusarium wilt. Research 2022:9818073

Tao C, Li R, Xiong W, Shen Z, Liu S, Wang B, Ruan Y, Geisen S, Shen Q, Kowalchuk GA (2020) Bio-organic fertilizers stimulate indigenous soil pseudomonas populations to enhance plant disease suppression. Microbiome 8:137

Tao C, Wang Z, Liu S, Lv N, Deng X, Xiong W, Shen Z, Zhang N, Geisen S, Li R, Shen Q, Kowalchuk GA (2023) Additive fungal interactions drive biocontrol of Fusarium wilt disease. New Phytol 238:1198–1214

Tian D, Song X, Li C, Zhou W, Qin L, Wei L, Di W, Huang S, Li B, Huang Q, Long S, He Z, Wei S (2021) Antifungal mechanism of Bacillus amyloliquefaciens strain GKT04 against Fusarium wilt revealed using genomic and transcriptomic analyses. MicrobiologyOpen 10:e1192

Trivedi P, Leach JE, Tringe SG, Sa T, Singh BK (2020) Plant–microbiome interactions: from community assembly to plant health. Nat Rev Microbiol 18:607–621

Wang B, Li R, Ruan Y, Ou Y, Zhao Y, Shen Q (2015) Pineapple–banana rotation reduced the amount of Fusarium oxysporum more than maize–banana rotation mainly through modulating fungal communities. Soil Biol Biochem 86:77–86

Wang B, Shen Z, Zhang F, Raza W, Yuan J, Huang R, Ruan Y, Li R, Shen Q (2016) Bacillus amyloliquefaciens strain W19 can promote growth and yield and suppress Fusarium wilt in Banana under greenhouse and field conditions. Pedosphere 26:733–744

Weller DM, Raaijmakers JM, Gardener BBM, Thomashow LS (2002) Microbial populations responsible for specific soil suppressiveness to plant pathogens. Annu Rev Phytopathol 40:309–348

Westphal A, Becker JO (2000) Transfer of biological soil suppressiveness against *Heterodera schachtii*. Phytopathology 90:401–406

Yang S, Shu C, Yang M, Zhou E (2019) *In vitro* toxicity test of soil disinfectants to *Fusarium oxysporum* f. sp. *cubence* and the control effects on Banana Fusarium wilt. Acta Agric Univ Jiangxiensis 41:249–257. (in Chinese)

Yang J, Duan Y, Liu X, Sun M, Wang Y, Liu M, Zhu Z, Shen Z, Gao W, Wang B, Chang C, Li R (2022) Reduction of banana fusarium wilt associated with soil microbiome reconstruction through green manure intercropping. Agric Ecosyst Environ 337:108065

Ye N, Wu Z, Gan H, Wang C, Mao W (2019) Studies on the control of *Fusarium oxysporum* f. sp. *cubense* by combinations of *Trichoderma* spp. formulations, China Fruits, pp 64–67. (in Chinese)

Yin L, Liu B, Wang H, Zhang Y, Wang S, Jiang F, Ren Y, Liu H, Liu C, Wan F, Wang H, Qian W, Fan W (2020) The rhizosphere microbiome of Mikania micrantha provides insight into adaptation and invasion. Front Microbiol 11:1462

Yuan X, Wang B, Hong S, Xiong W, Shen Z, Ruan Y, Li R, Shen Q, Dini-Andreote F (2021) Promoting soil microbial-mediated suppressiveness against Fusarium wilt disease by the enrichment of specific fungal taxa via crop rotation. Biol Fertil Soils 57:1137–1153

Zeng L, Lin W, Lv S, Wang F, Xia L, Liu W, Wu C, Zhou J, Du C, Cai K, Liu J (2019) Continual effect and soil microbial ecology mechanism of banana-sugarcane rotation controlling Fusarium wilt of banana(I). Chin J Eco-Agric. (in Chinese)

Zhou D, Jing T, Qi D, Feng R, Duan Y, Chen Y, Wang F, Zhang X, Xie J (2017) Isolation and identification of *Streptomyces lunalinharesii* and its control effect on the Banana *Fusarium* wilt disease. Acta Horticult Sin 44:664–674. (in Chinese)

Zhu J, Zhang G, Shang M, Liu H, Shen Z, Li R, Shen Q (2021) Isolation and identification of a thermostable *Trichoderma* strain for banana growth promotion in tropic area. Weishengwu Xuebao 61:206–218

Chapter 5
Field Application of *Pseudomonas fluorescens* for Rice Disease Management Improves Rice Yields and Leads to Development of Disease Suppressive Rice Soils in Southern India

P. Muralidharan, M. S. Rajeev, A. Reema, K. Sajnanath, G. Lekha, T. Sivakumar, and Samuel Gnanamanickam

1 Introduction

Gnanamanickam (2009) provided an overview of progress in biological rice disease control. In this overview, he narrated how this method of rice disease management started in his laboratory and described the advances that have been made. Areas and advances that were highlighted included, (i) new knowledge about the rice blast fungus, *Magnaporthe oryzae*, (ii) development of plant-associated *Pseudomonas* and *Bacillus* strains as inoculants for rice for the control of rice pathogens and plant growth-promotion/yield increase. This group of beneficial bacteria are the plant growth-promoting rhizobacteria (PGPR), and (iii) mechanisms of biological control used by PGPR for plant-growth promotion and rice disease suppression. It became known that the antifungal and antibacterial metabolites, such as 2,4-diacetylphloroglucinol, pyrrolnitrin and pyoluteorin were detected also in the Indian strains of these bacteria assembled from South Indian rice paddies (Velusamy et al. 2006).

Studies on the Development of Bacterial Biocontrol Agents for Rice Diseases. This work started in Gnanamanickam's laboratory at University of Madras in southern India back in 1982 and continued as intensive research over a 24-year period (1982–2006). Research carried out during this period can be summarized as the following:

P. Muralidharan · M. S. Rajeev · A. Reema · K. Sajnanath · G. Lekha · T. Sivakumar
Krishi Vigyan Kendra-Alappuzha, Indian Council of Agricultural Research (ICAR), Kayankulam, Kerala, India

S. Gnanamanickam (✉)
R&D, Biotech Innovations, Wylie, TX, USA
e-mail: sgnanamanickam@btilabs.com

S. Gnanamanickam, M. Höfte (eds.), *Disease Suppressive Tropical Soils*,
https://doi.org/10.1007/978-3-032-08100-1_5

1. Introduction of *Pseudomonas fluorescens* to India (Unnamalai and Gnanamanickam 1983).
2. Biological control of rice blast by seed treatment of rice with *Pseudomonas fluorescens* (carried out at the International Rice Research Institute in the Philippines) (Gnanamanickam and Mew 1989, 1992).
3. Genetic analysis of antibiotic production by *Pseudomonas fluorescens* Pf714 (Chatterjee et al. 1996; Valasubramanian 1994).
4. Identification of 2, 4-diacetylphloroglucinol produced by tropical strains of *P. fluorescens* as a major mechanism in the suppression of rice bacterial leaf blight caused by *Xanthomonas oryzae* pv. *oryzae* (Velusamy and Gnanamanickam 2003; Velusamy et al. 2006).
5. Role of 2,4-diacetyl phloroglucinol in rice sheath blight (*Rhizoctonia solani*) suppression (Immanuel 2006), and
6. Development of several *Pseudomonas fluorescens/P.putida* and *Bacillus* strains for the suppression of rice diseases (Table 5.1).

Starting about 1980 or earlier, perhaps there was a strong research focus on biocontrol of rice diseases at the International Rice Research Institute (IRRI) in the Philippines. T. W. Mew and his team of researchers were involved in using not only

Table 5.1 Bacterial biocontrol agents developed for major diseases of rice. Center for Advanced Studies in Botany, University of Madras, India

Disease	Pathogen (causal agent)	Biocontrol agent developed	Reference
Blast (Bl)	*Pyricularia grisea* (Teliomorph: *Magnaporthe oryzae*)	*Pseudomoans fluorescens* *Bacillus* spp.: *B. polymyxa, B. pumulus, B. coagulans, Enterobacter agglomerans*	Gnanamanickam and Mew (1992), Valasubramanian (1994), Kavitha (2002)
Sheath Blight (ShB)	*Rhizoctonia solani* (Teliomorph: *Thanetophorus cucumeris*	*P. fluorescens* *P. putida* *P. magaterium* *P. polymyxa* *P. pumulus* *P. coagulans,* *E. agglomerans*	Sakthivel and Gnanamanickam (1989), Thara (1994), Krishnamurthy and Gnanamanickam (1998), Kavitha (2002)
Sheath-rot (Sh-R)	*Sarocladium oryzae*	*P. fluorescens*	Sakthivel and Gnanamanickam (1987, 1989)
Bacterial Blight (BB)	*Xanthomonas oryzae* pv. *oryzae*	*Bacillus* spp. *B. lentus* *B. cereus* *B. circulans*, and *P. fluorescens*	Vasudevan (2002), Velusamy and Gnanamanickam (2003)

Source: Immanuel (2006)

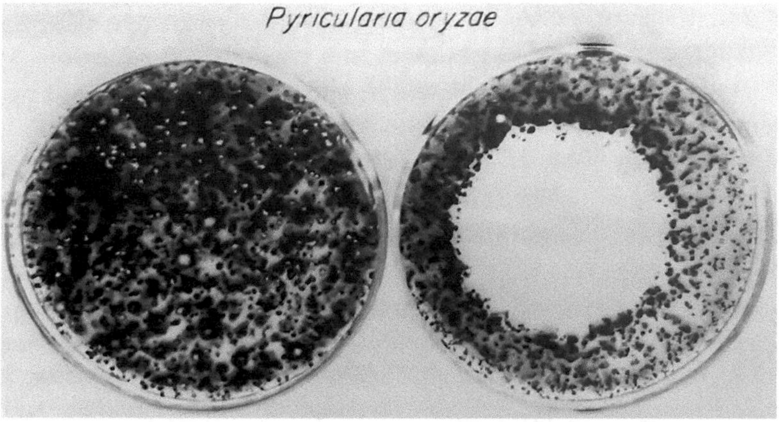

Fig. 5.1 Inhibition of *Pyricularia oryzae* in potato dextrose agar plate by fluorescent pseudomonad strain 7–14. The plate on the right was inoculated in the center with a loopful of cells of 7–14 after the agar surfaces of both plates were spread with conidial suspension of *P. oryzae*. The plate on the left (check) did not receive bacterial inoculation. Fungal inhibition was observed after 4 days of incubation. (Gnanamanickam and Mew 1992)

bacterial biocontrol agents (by bacterization) but also *Trichoderma* spp. (Mew and Rosales 1986; Rosales and Mew 1997; Gnanamanickam and Mew 1989, 1992) (Fig. 5.1). They significantly advanced the role of biological methods to manage rice diseases such as blast, sheath blight, bakane, and several others. Because of their coordinated effort in teaching and training rice researchers (starting from 1987 onwards) from rice growing countries of Asia, the message about the benefits derived from the use of environment-friendly and ecology-conscious biocontrol agents over chemicals/fungicides soon spread to other rice-growing countries of Asia. Countries like China, India, Malaysia, Thailand, Vietnam, and Indonesia started applying bacterial biocontrol agents (Fig. 5.2). Further, continuous monocropping of rice for long periods of time, most probably led to the development of disease suppressive rice soils similar to what has been observed in wheat as the take-all decline (TAD) (Raaijmakers and Weller 1998, 2001; Weller et al. 2002, 2007).

As an offshoot of the development of bacterial biocontrol agents, primarily from Gnanamanickam's laboratory, the biopesticide industry was born in India. This was marked by the manufacturing of *P. fluorescens* and *Bacillus* strains by small and big companies and agricultural universities. This led Kerala Agricultural University to make and distribute *P. fluorescens* to Kerala farmers. At this time, the field application of these biological alternatives in the Kuttanad rice production area of Kerala began in 2011. As a coordinated practice, it was a set of technology demonstrations put in place by the officials of the local Indian Council of Agricultural Research (ICAR) at Kayankulam, Kerala.

This set of technology demonstrations are described below in detail:

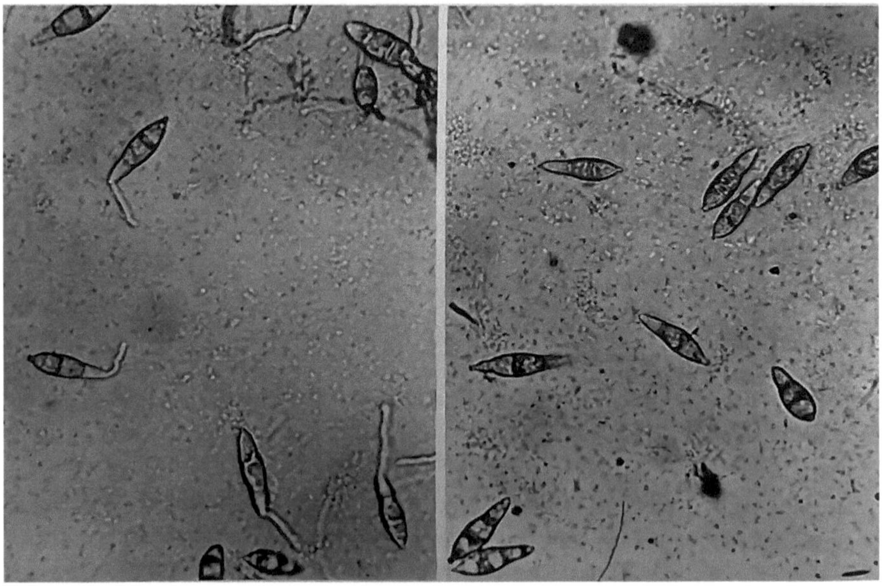

Fig. 5.2 Inhibition of conidial germination in Pyricularia oryzae by antiblast substances produced by *Pseudomonas* strain 7–14. Conidia were suspended either in water (left) or in an aqueous solution of AB, and AB2 at 1.0 ppm concentration (right). Germination counts were made after 6 hours. Antiblast substances inhibited conidial germination from 70% to 100%. (Gnanamanickam and Mew 1992)

Kuttanad in Kerala is exceptional because extensive paddy lands below the sea level, yet the area remains the state's principal rice-growing basin (Fig. 5.3). Most of this below-sea-level landscape lies in Alappuzha District and is recognized as an ecologically delicate agro-ecosystem. Chronic waterlogging, strongly acidic subsoils and erratic monsoon swings collectively heighten the system's fragility. Flood-induced crop losses are frequent in the low-lying padashekharam blocks, especially during summer cloudbursts and the main monsoon. Because of its distinctive landscape and long-standing rice-farming culture, FAO designated Kuttanad a "Globally Important Agricultural Heritage System" (GIAHS) in 2013. That recognition, however, also underscored the need to restore ecological balance and guide the waterlogged area toward more sustainable practices, as decades of intensive fertilizer and pesticide use have heightened environmental stress.

1.1 Pesticide Poisoning in India

This has been a major problem in India since at least 1958, when 100 people died after consuming flour contaminated with organophosphorus (OP) insecticide parathion (Gupta 2004). Figure 5.4 shows how an agricultural worker gets exposed to the chemical he applies to rice.

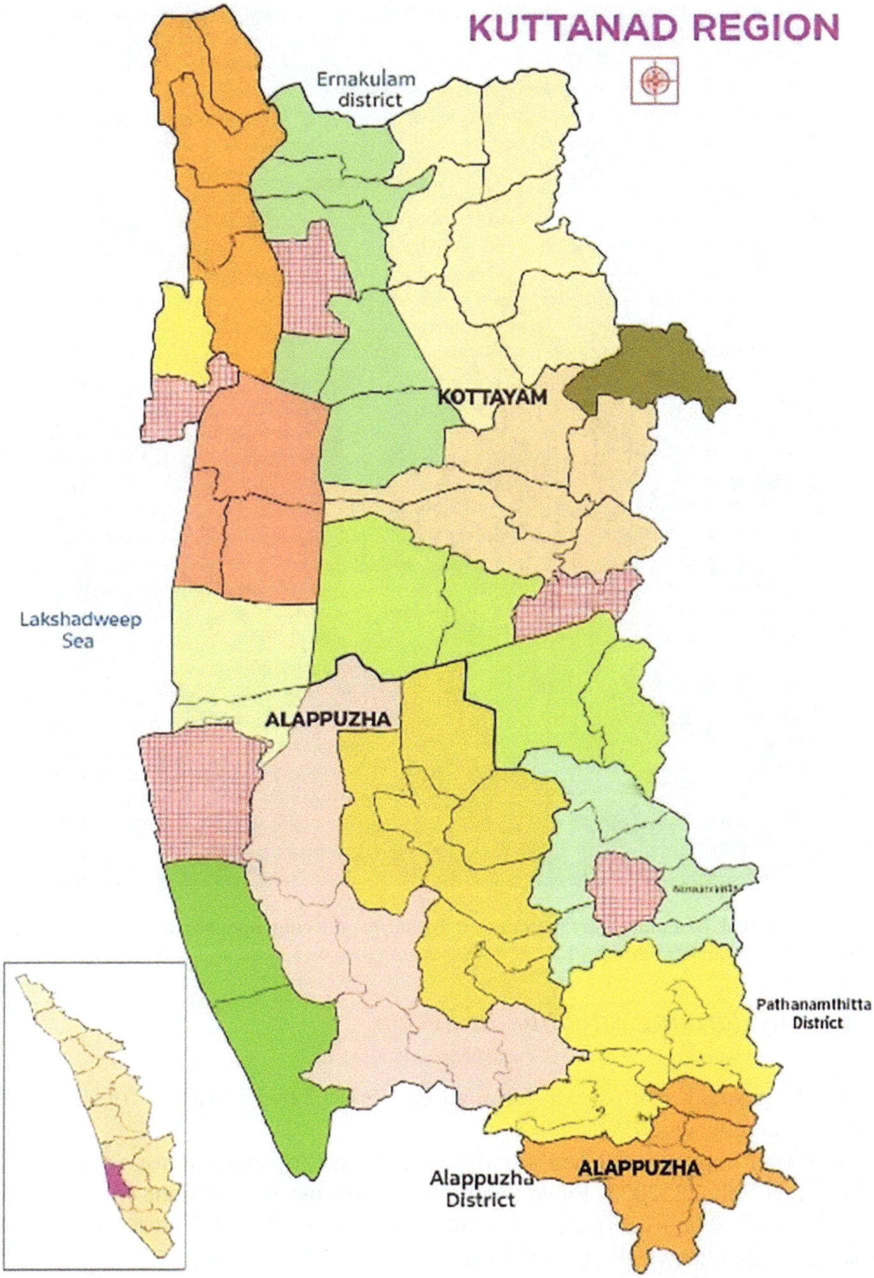

Fig. 5.3 (**a**) Administrative map of the Kuttanad Region, Kerala, India. Orange = Ernakulam district; green = Kottayam; yellow = Alappuzha; pink hatch = reclaimed polders below mean sea level. Thick black line = district boundary; dashed line = panchayat boundary. *(Created by S. Gnanamanickam).* (**b**) The Kuttanad region and Rice production in Kuttanad under Mean Sea Level (MSL). *(Photo by S. Gnanamanickam)*

Fig. 5.4 Typical pesticide-spray practice in Kuttanad showing operator exposure pathways. (*Photo by S. Gnanamanickam*)

1.2 An Alarming Situation in Kerala

In Kerala, pesticide overuse was turning Kerala's rice bowl into an ecological wasteland (Ameerudheen 2019).

A recent report published by the Department of Agriculture (Source ON Manorama, July 22, 2023) found that high levels of pesticide residue were found in samples of food items collected from the open market. The report was based on the results of laboratory testing carried out during Oct. 2022 to March 2023. Pesticide residues were detected in 242 out of 679 samples analyzed (35.64%) and these samples included vegetables, fruits, dried fruits and others.

In the backdrop of this alarming situation, the No Pesticide Rice Farming described in this chapter where rice farmers do not apply chemical pesticides and use much less fertilizers to manage the rice crops with biological alternatives (*Pseudomonas fluorescens* and trichocards (Kushwaha 2023) and still harvest

15–20% higher grain yields is very highly significant. This ecology-conscious practice is an affordable practice for resource-poor rice farmers while it is good for preserving rice soil health and human health.

To address all these problems, technology demonstrations were organized by the Krishi Vigyan Kendra (KVK) Alappuzha in Muttar Village of Veliyanad Block of Kerala during the year 2011–12. These demonstrations were conducted over four consecutive crop seasons (one season per year) as part of the National Innovations in Climate Resilient Agriculture (NICRA) initiative under the Indian Council of Agricultural Research.

2 Rice Cultivation Calendar Developed for Kuttanad

	Day	**Cultivation Operation**
Before Sowing	45th Day	Preparations for rice cultivation begin
	30th Day	Preparation of land by tractor/tiller
	25th Day	Application of dolomite/lime based on soil test (General recommendation: 140 kg dolomite/lime per acre)
	22nd Day	Wash off the fields with fresh water to allow weeds to germinate
	18th Day	Flood the fields for controlling the germinated weeds
	2nd Day	Seed Treatment: seed requirement for drum seeding is 12–15 kg/acre. Seeds are soaked in *Pseudomonas fluorescens* solution 10 g/ 5.0 ml for 1 kg seed.

	1st Day The day of sowing	Drain the solution and allow the seeds for germination. The fields are prepared for sowing by draining the water. Seeds are sown immediately after sprouting. If there is more moisture content, seeds are dried in shade. The sprouted seeds are filled into the seed drums up to half of their volume and the seeds are sown by pulling the drum seeder. A spacing of 40 cm is allowed after every 3 meters (2 lines of drum seeder).
After Sowing	**Day**	**Cultivation Operation Performed**
	7th Day	Flooding the fields with water
	14th Day	Fields are drained for fertilizer application
	15th Day	First fertilizer application (based on soil test result). General recommendation: 7 kg Urea + 45 g Factomphos + 10 kg Potash or 26 kg Urea + 100 kg Rajphos + 10 kg Potash/acre.
	16th Day	Broad-spectrum weedicide is applied
	18th Day	Fields are flooded with water

(continued)

	20th Day	A power weeder is used at weekly intervals (if herbicide application is not done)
	27th Day	Fields are drained and dolomite/lime is applied at 100 kg/acre.
	Up to 30th Day	Fields are monitored for thrips attack. Fields are submerged if there is thrips attack. Fiest application of tricho card (2 cc cards) to prevent stem borer and leaf folder.
	30–40 Days	Application of second dose of fertilizers (on the basis of soil test result)
	45th Day	Foliar application of 2% *Pseudomonas fluorescens* solution (5 ml/liter) against fungal diseases.
	55–60 Days	Application of third dose of fertilizers (25 kg Urea and 10 kg Potash)
	60th Day	Placing third batch of tricho cards.
	60–100 Days	Monitoring the field for BPH (brown plant hopper) attack.
	70th Day	Foliar application of soluble fertilizers (if third dose of fertilizers was not done) and adding 0.1% borax spray improves grain yield
	75th Day & onwards	Placing a fourth batch of tricho cards and monitoring fields for pest & diseases
	80th Day	Removal of wild rice and other weeds on bunds
	85th Day	Application of fish amino acid @4 ml/liter against rice bugs.
	105th Day	The fields are drained completely.
	115–120 day	Harvesting of the crop starts when 80% of the grains appear to have matured.

3 Identifying the Technology Package

Based on the problems identified through focus group discussions (FGD) with farmers and also with inputs from expert consultants, components of the technology package for demonstrations were developed (Muralidharan et al. 2017). The primary focus was to identify agricultural practices that could strengthen agro-ecosystem resilience in response to climate variability. A detailed calendar (shown in the previous section) for rice cultivation operations was developed. The package of technologies included highlighted the following:

1. <u>Optimization of seed rate and plant population</u> through the use of a drum seeder: A lightweight four-drum (Fig. 5.5.) seeder allows wet sowing of pre-sprouted rice seed onto puddled, levelled soil. The unit carries its four perforated seed drums on a single axle and rolls easily on side wheels, so one person can pull it across the puddled field. Each 9 mm aperture meters seed so that about three grains fall every 10 cm, with rows set 20 cm apart. Drums were filled halfway with water-soaked, pre-sprouted paddy rice seeds. On pulling the seeder, 3–4 paddy seeds are placed at 10 cm intervals at a row spacing of 20 cm and @ 30 kg/ha.

Fig. 5.5 Four-drum manually pulled seeder used for line sowing of pre-germinated rice seed in wet fields. (*Photo courtesy S. Gnanamanickam*)

2. Site specific acidity-nutrient management (SSNM) based on soil testing: Because Kuttanad soils are naturally strongly acidic, correcting pH is the starting point for improving nutrient-use efficiency and crop output. Soil analyses dictated the dose of dolomite, chosen both for its liming effect and for its contribution of magnesium. Fertilizer recommendations were adjusted based on soil test results, often resulting in lower application rates than those traditionally used by farmers.
3. Facilitation/Identifying partner farmers for demonstration. After village consultations, lead farmers who were willing to pilot the full package were recruited, and in the first year each volunteer signed a brief agreement outlining the responsibilities of hosting a demonstration plot. An application form was prepared, and farmers completed it, agreeing to carry out the specified demonstration activities.
4. From the second year onwards, a single padasekharam was selected based on willingness from the office bearers who had prior consent from the farmers of that padasekharam. These farmers individually submitted their application. A total of 114 farmers partnered in these demonstrations which covered an area of 74.2 ha in 4 years.
5. Training: At the start of every season KVK staff ran hands-on sessions covering each component of the package. A calendar of activities (shown in the previous section) was prepared with day zero being the day of sowing. The schedule included technical protocols covering pre-planting tasks such as seed treatments. Starting from the second year, these preparations became quicker, largely thanks to interactions among farmers. Key inputs, including dolomite, *Pseudomonas fluorescens* for seed treatments, and trichocards, were supplied to partner farmers through the third year. In the fourth year, dolomite was provided through State Department of Agriculture schemes, while the provision of *Pseudomonas* and trichocards continued as before.

6. Recording Data: Each participating farmer received a field notebook to document all farming operations from land preparation to harvesting, including quantities used and expenditures on inputs and labor.

In the first demonstration year (2011–12), a Farmer Field School was organized for all participating farmers to teach agro-ecosystem analysis and identification of pests and their natural enemies, as well as economic threshold levels for pests, diseases, and plant nutrition indicators. From the second year (2012–13) onward, the KVK team conducted regular field visits to individual plots (on rotation) up to harvest time. These visits facilitated thorough discussions about implemented technologies and allowed participants and visiting farmers to directly observe and learn about the impacts of these practices.

4 Major Outcomes

The overall achievements of the demonstrations for 4 years were summarized as follows:

- Using a paddy seeder (DRUM SEEDER) reduced the sowing rate to ~30 kg ha^{-1} (versus 100–120 kg ha^{-1} broadcast) and cut seed costs by roughly 25%. Line-sown seeds, spaced at 20 × 10 cm, provided optimal plant populations, resulting in more productive tillers and better crop aeration, significantly reducing pest and disease issues. The stronger root anchoring achieved by line sowing reduced lodging caused by wind and summer rains, compared to broadcast plots which typically experienced about 20% lodging losses.
- Thus, this technology enhanced the crop's resilience to climate stress. Moreover, harvesting time using a combined harvester was significantly reduced due to uniform plant populations and lower lodging, saving approximately 40–50% on harvesting costs (a lodged broadcast acre typically took around 90 minutes, while the drum-seeded plots required only about 45 minutes). Together, these improvements lowered overall cultivation costs by roughly 10–20%. Soil test guided dolomite (an anhydrous carbonate mineral composed of calcium magnesium carbonate (CaMg(CO3)2)) neutralized acidity, improved nutrient uptake and simultaneously supplied magnesium to the crop.
- Adjusting fertilizer amounts based on soil-test recommendations reduced fertilizer costs by about 30%, effectively covering the expense of dolomite. This site-specific nutrient management (SSNM) approach led to healthier crops and higher yields, while simultaneously reducing nutrient runoff and water pollution.
- Many farmers finished the season without resorting to synthetic pesticides.
- These plots stayed essentially pest-free, and farmers credited the combined use of *Pseudomonas fluorescens* and trichocards (Kushwaha 2023). In other instances, chemical pesticide and fungicide use dropped by as much as 90%. Farmers who previously depended heavily on chemical crop protection saw their pesticide and fungicide costs cut by approximately 50% compared to earlier sea-

sons. These changes lowered production costs and helped mitigate environmental pollution.

- Overall, this package of technologies delivered an average yield improvement ranging between 15 and 20%. While the traditional broadcast crops yielded an average of 5–6 tons/ha, by adopting this package of technologies, farmers could harvest 6–7 tons/ha with reduced input use. Throughout the four-year demonstration period, treated plots consistently showed increased productive tillers per square meter, more grains per panicle, and heavier grain weights. The higher yield obtained, and the reduced cost of cultivation led to AN OVERALL HIGHER NET PROFIT OF A MINIMUM OF RS.12,500 PER HA.
- FROM DEMONSTRATIONS TO WIDER IMPACTS: The demonstrations organized by the KVK impacted large number of farmers due to the support received from the following individuals and organizations:
- Effective KVK-ATMA Coordination: Results were shared at ATMA's (Agricultural Technology Management Agency) monthly advisory meetings, which sped up adoption in neighboring. Based on these experiences, ATMA organized numerous demonstrations across various blocks in Alappuzha district, focusing on drum seeder use, dolomite application, and bio-agent strategies for pest and disease control.
- Role of Partner Farmers: Participating farmers were satisfied and many willingly acted as master trainers, voluntarily disseminating the technology package to progressive growers in neighboring villages.

Role of DoA and Local Self-Government: In 2014–15 the Department of Agriculture's 50% subsidy scheme helped farmers source dolomite and use of *Pseudomonas fluorescens*-treated seeds in about 400 ha paddy fields in the project village and adjoining villages. The local self-government (Muttar Panchayath) also actively supported broader adoption of these practices by encouraging farmers to participate.

5 Conclusions

While the KVK demonstrations could convince several farmers to use a package of scientific practices and learn from their impacts (reduced use of inputs, less pollution, higher profits and enhanced climate resilience), the upscaling of the technology package was possible only through the concerted efforts of the Department of Agriculture, ATMA, Local Self Governments and input agencies. Apart from creating awareness on the merits of new technologies, farmers did hand-holding support while implementing the same and needed access to the recommended inputs (free initially but made available to buy later) at the right time. Our experience clearly shows the importance of engaging with other stakeholders such as ATMA, the State Department of Agriculture and the Local Self Government Institutions for the technology demonstrations to have much wider impacts.

Fig. 5.6 A rice farmer and officials of the Indian Council of Agricultural Research with Gnanamanickam, July 2022. *(Photo courtesy of S. Gnanamanickam)*

Pictures from Gnanamanickam's visit to KVK-Alappuzha on July 28, 2022, and December 2023 (Figs. 5.6 and 5.7). Rice farmers who use *Pseudomonas fluorescens* to protect their rice crop from diseases for 14 years are seen with the ICAR officials and Gnanamanickam who coordinated this No Pesticide Rice Farming package of practices.

6 Conclusion: Development of Disease Suppressive Rice Soils in Kerala

We repeat here again that rice production in Kerala has a very long history and perhaps has been cultivated in the same land for over hundreds and even thousands of years. The soil is acidic (or acid sulphate soil) and the application of *P. fluorescens*

Fig. 5.7 Gnanamanickam during a visit in December, 2023: collecting rhizosphere samples in farmer's fields. *(Photo courtesy of S. Gnanamanickam)*

leads to its buildup. *P.fluorescens* strains in the rice rhizosphere of India are known to produce the antibiotic, 2,4-diacetylphlorglucinol (DAPG) (Velusamy and Gnanamanickam 2003; Velusamy et al. 2006). Therefore, it is highly probable that the Kerala rice soils are disease suppressive. Careful monitoring is in progress to confirm this phenomenon (Figs. 5.8 and 5.9). DISEASE SUPPRESSIVENESS OF THESE KERALA RICE SOILS IS VERY SIGNIFICANT FOR PRESERVING THE RICE SOIL HEALTH. IT ALSO PREVENTS POLLUTION OF THE WATER BODIES CAUSED BY THE PRESENT PRACTICE OF EXCESSIVE USE OF CHEMICALS AND PESTICIDES.

Therefore, we believe that the No Pesticide Rice Farming is an ecology-conscious and environment-friendly rice production practice that can be afforded by even resource-poor rice farmers of India and Kerala (Fig. 5.10).

Gnanamanickam has collaborated with RARS PATTAMBI scientists for more than 20 years (Fig. 5.11).

Therefore, we the authors very strongly believe that in areas of the country where rice has been grown for hundreds and thousands of years, such as the Kuttanad rice growing region of Kerala, application of *P.fluorescens* (that produced the antibiotic 2,4diacetylphloroglucinol (DAPG) (Velusamy and Gnanamanickam 2003; Velusamy et al. 2006) or PGPR strains by rice farmers have led to soils that have progressively become suppressive to rice blast, sheath blight, bacterial leaf blight and other major rice diseases.

Fig. 5.8 Antibiotics produced by *Pseudomonas fluorescens* cause pathogen suppression and plant growth promotion/yield increase in rice. (*Diagram by S. Gnanamanickam)*

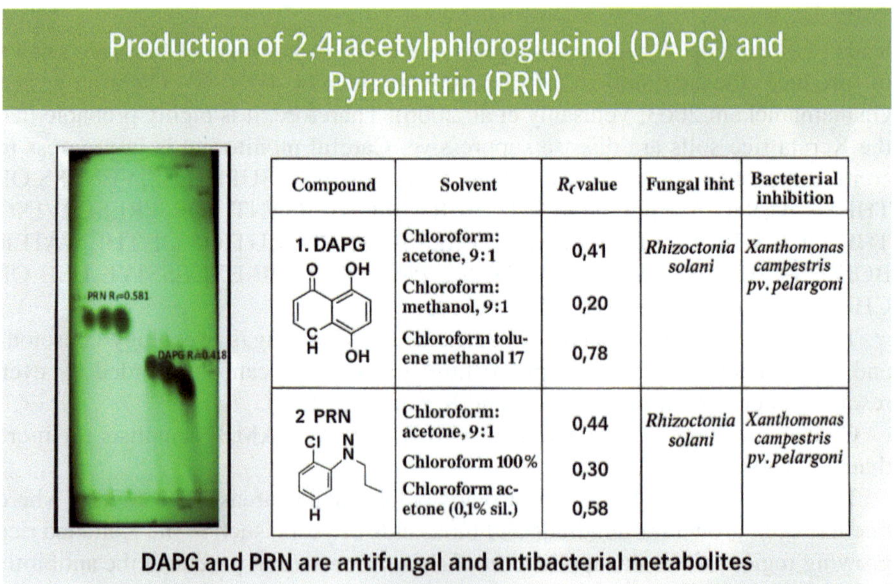

Compound	Solvent	R_f value	Fungal ihht	Bacteterial inhibition
1. DAPG	Chloroform: acetone, 9:1 Chloroform: methanol, 9:1 Chloroform toluene methanol 17	0,41 0,20 0,78	*Rhizoctonia solani*	*Xanthomonas campestris pv. pelargoni*
2 PRN	Chloroform: acetone, 9:1 Chloroform 100% Chloroform acetone (0,1% sil.)	0,44 0,30 0,58	*Rhizoctonia solani*	*Xanthomonas campestris pv. pelargoni*

Fig. 5.9 Farmers and ICAR-KVK staff showcase a pesticide-free rice harvest from a demonstration plot in the Kuttanad region of Kerala. (*Photo courtesy of S. Gnanamanickam, 2012)*

Fig. 5.10 Rice planting operation at the Regional Agricultural Research Station, Pattambi, Kerala, India. (Pictures from Gnanamanickam's visit on July 31, 2022)

Fig. 5.11 Most of the greenhouse and field experiments described in his recent book (Gnanamanickam 2023) were carried out in Pattambi, Kerala (*Photo courtesy of S. Gnanamanickam*)

And what is reported and witnessed from the recent visit of Gnanamanickam to leading rice farmers of the Kuttanad rice-growing region in Kerala is that they have used *P. fluorescens* for the last 14 years to protect their rice crop from rice diseases and have had increased rice productivity by 15–20%. These farmers have harvested disease-free crops without applying any plant protection chemicals. This strongly suggests that these soils, of acidic/acid-sulfate nature, due to flooding may have become suppressive to rice diseases. We would like to carry out some careful monitoring of the buildup of *P. fluorescens* in these soils to draw this conclusion.

6.1 Disease Suppressive Soils: Development

Disease-suppressive soils develop through a combination of inherent soil characteristics and external environmental influences. These include physical, chemical, and biological aspects of soil health, as well as factors such as land use changes, climate variability, and agricultural practices like nutrient or water management (Jayaraman et al. 2021).

General process needs to be followed to study specific disease-suppressive soils. Only when this is accomplished will we know for sure if the Kuttanad rice field soils are suppressive to blast and major rice pathogens.

6.2 Rice Blast Disease

In this recent book on Rice Blast Disease: Contributions to Biological Management and the Bio-pesticide Industry (2023) (Fig. 5.12), Gnanamanickam has tried to bring an update both on the status of Rice Blast disease in India and Asia and an update on the use of biofungicides, in particular, *Pseudomonas fluorescens* by the rice farmers.

Fig. 5.12 The concept that has strongly emerged is Disease Suppressive Rice Soils and the theme/concept became an important feature of this new volume on Rice Blast Disease. (Copyright Gnanamanickam 2023)

The severity of the Rice Blast appears to stay as a continuing threat to rice cultivation in India and Asia. The pathogen is characterized by its gray-colored, oval-shaped colony morphology and distinct microscopic features. Especially concerning is the emergence of Wheat Blast in Bangladesh, which poses a renewed threat to wheat production in these areas.

The update on the use of biofungicides or biopesticides by rice farmers is amazingly of a larger volume in India. Judging from the number of manufacturers who produce *P.fluorescens* in different forms, names, volumes and prices, India leads the way and uses its India Mart platform to export to several other countries including the United States. There are challenges to the growth of this industry, and Gnanamanickam (2023) was able to outline only some of them.

Other nations of Asia do use *P. fluorescens* and biopesticides in rice farming, but the growth of the industry is limited.

References

Ameerudheen TA (2019) in *Newsletters Scroll* in

Chatterjee A, Valasubramanian R, Vachani A, Ma W-L, Gnanamanickam SS, Chatterjee AK (1996) Biocontrol of rice diseases with *Pseudomonas fluorescens* 7–14: isolation of ant mutants, cloning of ant[+] DNA and an evaluation of a role for antibiotic production in the control of blast and sheath blight. Biol Control 2:185195

Gnanamanickam SS (2009) Biological control of rice diseases. Springer Science+Business Media B. V, 108p

Gnanamanickam SS (2023) Rice blast disease: contributions to biological disease management and the biopesticide industry. Copyright 2023. S. S. Gnanamanickam, 222p

Gnanamanickam SS, Mew TW (1989) Biological control of rice blast with antagonistic bacteria. Int Rice Res Newslett 14:34–35

Gnanamanickam SS, Mew TW (1992) Biological control of blast disease of rice (*Oryza sativa*, L) with antagonistic bacteria and its mediation by a *Pseudomonas* antibiotic. Ann Phytopathol Soc Jpn 58:380–385

Gupta P (2004) Pesticide exposure-India scene. Toxicology 198:83–90

Immanuel JE (2006) Molecular analysis of 2, 4-diacetylphloroglucinol (DAPG) production by strains of Pseudomonas fluorescens in rice rhizosphere and assessment of its role in suppression of sheath blight (ShB) disease of rice. Ph. D Dissertation, University of Madras

Jayaraman S, Naorem AK, Lal R, Dalal RC, Sinha NK, Patra AK, Chaudhari SK (2021) Disease-suppressive soils-beyond food production: a critical review. J Soil Sci Plant Nutr 21:1437–1465. https://doi.org/10.1007/s42729-02100451-x. Epub 2021 Mar 12. PMID: 33746349; PMCID: PMC7953945

Kavitha S (2002) Strategies for management of rice blast and sheath blight with bacterial biocontrol agents in combination with major genes for disease resistance. Ph. D Dissertation, University of Madras

Krishnamurthy K, Gnanamanickam SS (1998) Biological control of rice blast by *Pseudomonas fluorescens* strain 7-14: evaluation of a marker gene and formulations. Biol Control 13:158–165

Kushwaha P (2023) Tricho cards preparation technology and its role in biological control. Bio Res Today 5:461–464

Mew TW, Rosales AM (1986) Bacterization of rice plants for control of sheath blight caused by *Rhizoctonia solani*. Phytopathology 76:1260–1264

Muralidharan P, Rajeev MS, Anand R, Sajnanath K, Lekha G, Sivakumar T (2017) Resource-conserving and eco-friendly technologies for climate resilience and higher profitability of rice production in Kuttanad region of Kerala. ICAR-KVKAlappuzha, ICAR-CPCRI, Regional Station, Kayankulam, 12p

Raaijmakers J, Weller DM (1998) Natural plant protection by 2,4diacetylphloroglucinol-producing *Pseudomonas* spp. in take-all decline soils. Mol Plant-Microbe Interact 11:144–152

Raaijmakers J, Weller DM (2001) Exploiting diversity of 2,4diacetylphloroglucinol-producing *Pseudomonas* spp.: characterization of superior rootcolonizing *P. fluorescens* strain Q8r1-96. Appl Environ Microbiol 67:2545–2554

Rosales AM, Mew TW (1997) Suppression of *Fusarium moniliforme* in rice by rice-associated antagonistic bacteria. Plant Dis 81:49–52

Sakthivel N, Gnanamanickam SS (1987) Evaluation of *Pseudomonas fluorescens* for suppression of sheath-rot disease and enhancement of grain yields in rice, *Oryza sativa*, L. Appl Environ Microbiol 53:2056–2059

Sakthivel N, Gnanamanickam SS (1989) Incidence of different biovars of *Pseudomonas fluorescens* in flooded rice rhizospheres in India. Agriculture, Ecosystem and Environment 25:287–298

Thara KV (1994) Biological control of rice sheath blight by bacterial antagonists: mechanisms of disease suppression. Ph. D Dissertation, University of Madras

Unnamalai N, Gnanamanickam SS (1983) *Pseudomonas fluorescens* is an antagonist to *Xanthomonas citri* (Hasse) dye, the incitant of citrus canker. Curr Sci 50:703–704

Valasubramanian R (1994) Biological control of rice blast with Pseudomonas fluorescens Migula: role of antifungal antibiotics in disease suppression. Ph. D Dissertation, University of Madras

Vasudevan P (2002) Isolation and characterization of Bacillus spp. from the rice rhizhosphere and their role in biological control of bacterial blight of rice caused by *Xanthomonas oryzae pv. oryzae*. Ph. D Dissertation, University of Madras

Velusamy P, Gnanamanickam SS (2003) Identification of 2,4-diaceylphloroglucinol (DAPG) production by plant-associated bacteria and its role in suppression of rice bacterial blight in India. Curr Sci 85:1270–1273

Velusamy P, Immanuel JE, Gnanamanickam SS, Thomashow LS (2006) Biological control of rice bacterial blight by plant-associated bacteria producing 2, 4diacetrylphloroglucinol. Can J Microbiol 52:56–65

Weller DM, Raaijmakers JM, Gardener BBM, Thomashow LS (2002) Microbial populations responsible for specific soil suppressiveness to plant pathogens. Annu Rev Phytopathol 40:309–348

Weller DM, Landa BB, Mavrodi OV, Schroeder KL, De La Fuente L, Blouin Bankhead S, Allende Molar R, Bonsall RF, Mavrodi DV, Thomashow LS (2007) Role of 2,4-diacetylpholoroglucin ol-producing fluorescent *Pseudomonas* spp. In the defense of plant roots. Plant Biol (Stuttgart, Germany). 9:4–20

Chapter 6
Suppressive Soils for the Management of Basal Stem Rot Disease of Oil Palm

Adeline Su Yien Ting

1 Introduction

The oil palm industry is a lucrative industry in the tropics. Many of these oil palm producing countries rely on the oil palm *Elaeis guineensis* Jacq. as an important economic cash crop. Oil palm produces raw materials (palm oil) that can be refined for various applications (industries, cosmetic, pharmaceutical, food, biofuel). This led to a great demand for the palm oil produced and their refined products. The growth of the oil palm industry is however, impacted by the threat of oil palm diseases. Of the many diseases, the Basal Stem Rot (BSR) disease, caused by *Ganoderma boninense*, is a major limiting factor to the industry. This disease causes the infected palm trees to rot at the base of the stem. The roots become infected and yellowing of fronds can be observed (Rees et al. 2012; Corley and Tinker 2003). In later stages, the fruiting body of the pathogen appears at the base of the stem. This often indicates that the basal stem is hollow, causing the palm tree to collapse (Lisnawita and Tantawi 2016). The BSR disease can cause significant losses to the oil palm industry by reducing yield and shortening their economic lifespan, rendering losses estimated at USD 500 million annually (Woittiez et al. 2017; MPOB 2019).

Several cultural, physical, chemical, and biological based control measures have been introduced over the years, with varying results. One interesting approach is the use of suppressive soil that has garnered interest due to its sustainable and environmentally friendly approach. Suppressive soils are soils that are able to inhibit the growth of the pathogen, even in the presence of a susceptible host and a virulent pathogen. Naturally occurring suppressive soils are well-described for several types of soil that have the ability to suppress Fusarium wilt, *Rhizoctonia, Streptomyces*

A. S. Y. Ting (✉)
School of Science, Monash University Malaysia,
Bandar Sunway, Selangor Darul Ehsan, Malaysia
e-mail: adeline.ting@monash.edu

S. Gnanamanickam, M. Höfte (eds.), *Disease Suppressive Tropical Soils*,
https://doi.org/10.1007/978-3-032-08100-1_6

scabies and *Gaeumannomyces tritici.* Some soils are augmented to induce disease suppression, that is by manipulating their biotic or abiotic components. Suppressive soils can be artificially induced by ensuring they are generally rich in beneficial microbial communities, and have ideal soil properties (i.e., high levels of organic matter and soil nutrients, good soil texture, pH, moisture levels).

However, suppressive soils specific for oil palms are not well studied, especially naturally occurring suppressive soils. To date, only the Blenheim soil in Malaysia is presumed to demonstrate disease suppression towards the BSR disease. Efforts to replicate disease suppressive soils by artificially inducing suppressive soils have been attempted, with varying degree of success. This chapter will elaborate on suppressive soils for oil palm and some of the challenges and future prospects of adopting the suppressive soil approach for management of BSR disease of oil palm.

2 The Oil Palm Industry

The oil palm industry flourishes in countries spread throughout the Southeast Asia, Africa and Latin America, due to the tropical climate (Liew et al. 2015). In the Southeast Asian region, Indonesia and Malaysia are the leading nations of oil palm cultivation and palm oil production, contributing to 85% of the world's palm oil production (Sime Darby 2014). Indonesia is currently the world's leading producer and exporter of palm oil, with a plantation acreage of 12.3 million hectares (ha) (censused in 2017) (Sigit et al. 2016). Malaysia has a humbler acreage, with 5.85 million ha of oil palm plantations (censused in 2018) (Kalam and Masjuki 2002). Nevertheless, the increase in land acreage in Malaysia, from 1.5 million ha in the year 1985 to 5.74 million ha in 2016 to 5.85 million ha in 2018, underpins the importance of the oil palm industry as the main economic growth generator for Malaysia (MPOB 2017; Oil Palm Planted Area 2018). The other palm oil producing countries (e.g., Thailand, Papua New Guinea, Colombia, Ecuador, Nigeria) has a cumulative 19 million ha of oil palm plantations (0.36% of the world's agricultural land in total oil palm planted area) across the tropics (Kushairi et al. 2018).

The oil palm tree originates from the family Arecaceae (formerly Palmae), and the subfamily Cocosoideae (Corley and Tinker 2003). It is a perennial crop, which allows the palm trees to continuously bear fruit all year and a capacity to produce palm oil for an average of 25 years in the field. Oil palm is a highly desirable oil crop with high productivity and versatility. The productivity of oil palm is 3–8 times more compared to other oil crops (e.g., rapeseed, sunflower, soy, canola) (Barcelos et al. 2015), contributing to 36% of the total worldwide vegetable oil production (Sambanthamurthi et al. 2000). Productivity is also considered economical, feasible and profitable, as the oil palm trees generally have higher output while commanding lower production cost (Dislich et al. 2017). Oil palm also has greater versatility. The palm oil derived has a broad range of oleochemical applications; from food industries to cosmetics, pharmaceuticals, and for biodiesel usage as well (Corley and Tinker 2003; MPOB 2017).

The production of palm oil is expected to address the upsurge of global demand for vegetable oils and biofuels in the years to come, estimated at 240 Mt and 277 Mt, respectively (Koh 2007; Corley 2009). With a healthy growth, the palm oil industry is expected to grow continuously as a global industry. The industry bloomed from a value of USD 65.73 billion in 2015, to USD 92.84 billion in 2021, with signs of continued upward trajectory (USDA 2012; Bentivoglio et al. 2018; Fathana 2018). This is also fuelled by the increasing demand for palm oil by countries in Asia (e.g., China, India, Pakistan, Bangladesh, the Philippines), the European Union (EU), the United States (USA), and Africa (e.g., Nigeria) (Sime Darby 2014; Kushairi et al. 2018). Additionally, the price for palm oil is the lowest among other vegetable oils, thereby increasing their demand (USDAFAS 2018).

This lucrative industry is unfortunately, threaten by disease infection to the oil palm trees. Diseases such as basal stem rot (by *Ganoderma boninense*) (Idris et al. 2011), vascular wilt (by *Fusarium oxysporum* f. sp. *elaeidis*) (Tengoua and Bakoum 2008), red ring disease (by *Rhadinaphelenchus cocophilus*) (Carlos and Ricardo 2004), sudden wilt (by *Phytomonas staheli*) and spear rot (by an unknown pathogen) (Corley and Tinker 2003), are common in oil palm plantations. Certain diseases are more prevalent in certain countries. For example, the bud rot disease of the fresh fruit bunch caused by *Phytophthora palmivora* is prevalent in Colombia and Ecuador (Martinez et al. 2018). In the African regions, Fusarium wilt disease is more prevalent (Paterson 2022). In Malaysia, the Basal Stem Rot (BSR) disease is dominant. The BSR disease can cause severe losses to many plantations. Darmono (2000) projected (based on 1996 prices) that a 1% disease incidence is translated to an annual loss of USD 38 M in Indonesia (Darmono 2000), while in Malaysia, losses due to BSR disease is estimated to be as high as USD 500 million annually (Woittiez et al. 2017; MPOB 2019). Therefore, it is critical that oil palm diseases are understood and managed in a sustainable manner, so that continuous production can be achieved.

3 Basal Stem Rot Disease of the Oil Palm

The oil palm trees are susceptible to disease causing bacterial and fungal pathogens. The most severe disease recorded in Southeast Asia is the Basal Stem Rot (BSR) disease caused by the pathogen *Ganoderma boninense* (Idris et al. 2011; Purnama et al. 2012) (Fig. 6.1). This disease is the most prevalent compared to other diseases affecting the oil palm trees, such as vascular wilt disease (Tengoua and Bakoum 2008), bud and spear rot (Corley and Tinker 2003), sudden wilt (Maluin et al. 2020), and red ring disease (Maluin et al. 2020), among others. The BSR disease, or also known as the Ganoderma white rot disease, is also the major disease affecting oil palm trees globally. Infected palms typically experience rotting at the lower part of the palm tree (i.e., base of the stem, root system), while the upper part of the palm trees shows yellowing of fronds (Rees et al. 2012; Corley and Tinker 2003). The fruiting body gradually appears at the base of the palm tree, and this often indicates

Fig. 6.1 A 10-day old culture of *Ganoderma boninense* on Malt Extract Agar

that the basal stem is hollow which may subsequently lead to palm trees collapsing (Lisnawita and Tantawi 2016). Consequently, infected palms have lower yield and shorter economic lifespan (Corley and Tinker 2003; Kushairi et al. 2019).

The BSR disease was first discovered in the Republic of Congo (Wakefield 1920). Over the years, 15 species of *Ganoderma* have been associated with BSR disease which includes *G. boninense, G. miniatocinctum, G. zonatum, G. xylonoides, G. tornatum,* and *G. chalceum* (Steyaert 1980; Turner 1981; Zhou et al. 2015). Comparatively, BSR disease incidence is higher in both Indonesia and Malaysia compared to other countries such as Africa, Papua New Guinea, and the neighbouring Thailand (Idris et al. 2004a). It has been hypothesized that this is attributed to the type of cropping system implemented and the type of soil used in these two countries. Monocropping of oil palm trees in coastal areas that were previously planted with coconut trees has led to 80% infection and 50% yield loss (Lim et al. 1992). For soil types, peat soils gave rise to palm trees with higher disease incidences as well (Rao et al. 2003). Replanting also often results in high disease incidence where oil palms replanted in same soils after 5 years can record 40–50% disease incidence, that increases to 65–75% in the next 7 years (Subagio and Foster 2003).

4 Control Measures for BSR Disease

Various control measures have been adopted for the control of BSR disease. Common approaches include cultural, physical and chemical methods. The cultural and physical methods are one of the most fundamental methods applied and they are usually implemented after a field assessment (Fig. 6.2). This typically involves soil mounding at the base of the stem, trenching, ploughing and harrowing, as well as removal of basidiocarps to prevent spread of the disease (Turner 1981). In some cases, felling and chipping of infected oil palm trees are carried out. Additionally, when infected root boles are detected, deboling of the infected root bole can be performed (Hushiarian et al. 2013).

For chemical methods, several fungicides have shown potential to control the disease. In general laboratory tests, the fungicides carbendazim (EC_{50} of 0.026 g/mL), hexaconazole (EC_{50} of 0.026 g/mL), pyraclostrobin (EC_{50} of 0.25 g/mL) and azoxystrobin (EC_{50} of 0.53 g/mL) have been found to be able to inhibit the pathogen

Fig. 6.2 Assessments to oil palm trees in the plantations are conducted periodically and interventions are implemented as required

(Ramasamy 2000; Idris et al. 2010; Said et al. 2019). In the field, systemic fungicides such as bromuconazole, hexaconazole, tetraconazole and triademenol, applied via trunk injection, has been found to be reasonably effective where disease progression is delayed (Chung 1990; Lim et al. 1990). Palms treated with hexaconazole reportedly had 74.4% survival rate, and prolonged fruit bunch production for another 5 years (Idris et al. 2004b). Nevertheless, the application of fungicides via trunk injection or drenching reportedly causes environmental and toxicity concerns.

Both the cultural-physical and chemical control methods have not met with much success, prompting investigations into the use of biological approach. The biological control of BSR disease has been attempted by many, with most biocontrol agents derived from soil, rhizosphere, endophytic communities, or other sources. Among these, the rhizosphere bacteria are widely studied for their biocontrol attributes (Gomez et al. 2017) with *Bacillus* spp., *Burkholderia* spp., *Pseudomonas* spp., and *Streptomyces* spp. reportedly able to inhibit *G. boninense* (Susanto et al. 2005; Irma et al. 2018; Lim et al. 2018; Shariffah-Muzaimah et al. 2020; Sujarit et al. 2020). These rhizosphere bacteria were able to confer disease control at the laboratory or glasshouse stage, but the control efficacy was not replicated in the field. One hypothesis for the inferior control effects observed in the field is that disease suppression is most effectively conferred by microbial communities rather than by one specific species (Gomez et al. 2017). As such, introduction of single microbial inoculants may not confer the expected results. Disease suppression is also influenced by soil physicochemical properties (i.e., soil organic matter, soil pH) (De Corato 2020b; Anothai and Chairin 2020). With these revelations, it is clear that disease suppression is influenced by both the biotic and abiotic components of the soil.

5 Suppressive Soil

Soil is a matrix that host a complex–dynamic ecosystem of diverse microbiome (Guerra et al. 2020). This includes a variety of bacteria, fungi, archaea, algae, protozoa, and nematodes (Guerra et al. 2020). Of the various types of soils, disease suppressive soils are highly favoured as they have been found to result in low disease incidence. In these disease suppressive soils, disease development is low, although susceptible hosts and virulent pathogens are present (Mazzola 2002; Siegel-Hertz et al. 2018). Baker and Cook (1974) fittingly described suppressive soils as "soils which the pathogen does not establish or persist, or establishes but causes little or no damage, or establishes and causes disease for a while but thereafter the disease is less important although the pathogen may persist in the soil". It is theorized that suppressive soils have highly desirable microbial composition and the activities of these microbes may lead to low disease incidence. The microbial communities may directly inhibit the pathogen, causing low pathogen inoculum or poor expression of disease infection, leading to low disease incidence. Furthermore, a soil is said to be disease suppressive when continuous cropping of the same

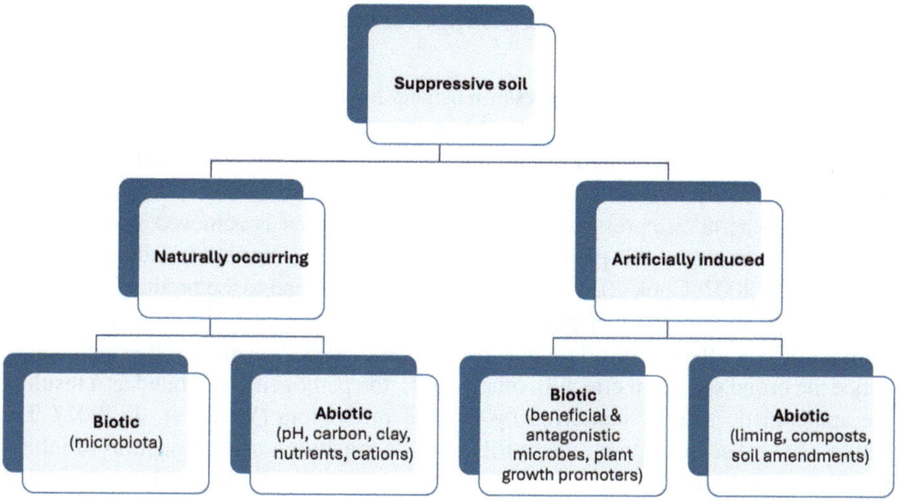

Fig. 6.3 Key biotic and abiotic components to naturally and artificially induced suppressive soils

susceptible host plant in the said soil, demonstrated low disease incidence even in the presence of pathogens in the soil (Weller et al. 2002; Schlatter et al. 2017).

Disease suppression by suppressive soils is attributed to both the biotic and abiotic elements in the soil (Fig. 6.3). The biotic elements are the soil microbial communities, while the abiotic elements include soil physical and chemical properties (Expósito et al. 2017). In suppressive soils, the biotic elements are pertinent in conferring disease suppression. The functions of the biotic components can either be eliminated via sterilization, or be transferred when sufficient suppressive soil is transferred to conducive soil (Stutz et al. 1986). The soil microbiome is evidently the key to achieving successful disease suppression. Suppressive soils are therefore, widely acknowledged as a sustainable disease management approach. With suppressive soils, disease suppression can be achieved with less chemical inputs required. As such, suppressive soil promotes sustainable food production, with high crop production anticipated with low generation of environmental footprints.

6 Naturally Occurring Suppressive Soils

While disease suppressive soils are highly desirable, there are only a few naturally occurring suppressive soils that have been described to date (Kyselkova et al. 2009). These naturally occurring suppressive soils are capable of inhibiting pathogen growth, stimulating plant defence mechanisms, and promoting plant growth for the benefit of the plants. Suppressive soils typically confer either the 'general' or 'specific' disease suppression, and they are not mutually exclusive and often co-exist (Jambhulkar et al. 2015; de Corato 2020b).

6.1 'General' and 'Specific' Suppressive Soils

Soils with **'general' disease suppression** usually have a broad-spectrum effect towards pathogens, but are unfortunately, not able to transfer the disease suppression to new soils or new fields. The limited potential of 'general' suppressive soils is that they suppress pathogens by expressing non-specific antagonistic activities towards the pathogens. For 'general' suppressive soils, disease suppression is achieved when the soil microbiota competes with pathogens for resources (e.g., nutrients) and/or for space (Weller et al. 2002; Cook 2014). Their interactions may lead to the production of compounds (e.g. antibiotics) that inhibit the pathogen (Hadar and Papadopoulou 2012). This inhibitory effect may affect not just the targeted pathogen but other pathogens, hence the broad spectrum effect. In other words, the pathogen is inhibited as a result of the antagonistic activity conferred by the soil microbiota (Weller et al. 2002). The nature of the disease suppression, attributed to competition and production of inhibitory compounds such as antibiotics, is complex and may not be easily replicated in other soils (de Corato 2020a). In fact, 'general' suppression is said to originate from the natural pre-existing elements in the suppressive soil. Therefore, the efficacy of these 'general' suppressive soils are dependent on field practices, whereby introduction of soil amendments (e.g. organic matter) or antagonistic isolates (e.g. Streptomycetes) are known to enhance disease suppression (Cook 2014; Tomihama et al. 2016).

By contrast, soils with **'specific' disease suppression** are able to inhibit pathogens and persist through future cycles, even in the absence of the susceptible host plants (Schlatter et al. 2017). This has been observed for monoculture cropping systems where disease incidence remained low even in the subsequent planting cycles. The 'specific' disease suppression is attributed to the function of individual or specific groups of beneficial microbes that inhibits the growth of the pathogen (Cook 2014). And because pathogen inhibition is conferred by a specific group or individual isolates, this disease suppression effect can be transferred from soil to soil and from one field to another, including the transfer to conducive soil (Andrade et al. 1994). 'Specific' disease suppression is therefore, highly effective, and can be achieved by transferring pure cultures or small portions (1–10%) of the suppressive soil to conducive soil (Weller et al. 2002). The mechanisms of disease suppression demonstrated by the isolates in soils with 'specific' disease suppression include parasitism and predation (Chandrashekara et al. 2012).

It has been postulated that suppressiveness functions as a continuum from 'general' to 'specific', with the 'general' suppression giving rise to 'specific' suppression over time. Soils with 'specific' disease suppression can either develop 'long-standing type of disease suppression', or be developed into soils with 'induced' disease suppression. 'Long-standing type of disease suppression' is developed when disease suppression remained with the soil even in the absence of the susceptible host plants (Weller et al. 2002). The suppressive soils from the Chateaurenard region of France are known to have 'long-standing disease suppression' towards Fusarium wilt disease (Alabouvette 1986). For 'induced' disease suppression, this can be achieved by sustaining disease suppression via monoculture cropping, inoculating pathogens into soil, or by continuously cultivating the same susceptible crops in the same soil (Weller et al. 2002). One example of 'induced' suppressive soil is the soil suppressing the take-all disease of wheat, whereby the continuous monocropping of wheat in the suppressive soil kept the disease incidence low throughout the many cycles (Weller et al. 2002). As 'specific' disease suppression is attributed to a few microbial taxa and their mechanisms of inhibition (i.e., parasitism, predation, induced host resistance), it is therefore easier to transfer the disease suppression to conducive soils.

6.2 Components to Suppressive Soils

It is therefore, evident that **soil microbiota** is the key to suppressive soil (Fig. 6.3). The microbiota confer disease suppression via mechanisms that promote plant growth as well as through the direct inhibition of pathogen. The high microbial diversity and species richness enhance the functions of the soil microbiota to contribute towards disease suppression, while promoting the survival of the plants (Janvier et al. 2007; Wei et al. 2019). In recent times, high-throughput sequencing has enabled the deciphering of the various microbial diversity found in soils suppressive to pathogenic *Fusarium* spp. (Siegel-Hertz et al. 2018), *Pythium irregulare* (Ros et al. 2020) and *Rhizoctonia solani* (Mendes et al. 2011). In these suppressive soils, the fungal and bacterial genera with known antibiotic activities are generally more abundant (Siegel-Hertz et al. 2018; Ros et al. 2020). This include Actinobacteria, Firmicutes and Proteobacteria (Mendes et al. 2011). The microbiome analysis of soils related to suppression of BSR disease in oil palm is however, not well explored. To date, most soil microbiome studies related to oil palm ecosystems were on effects of land use. Their microbiome was revealed through 16S- and 18S-targeted amplicon sequencing to profile the different communities found in soils used for oil palm cultivation, rubber planting, rainforests, and logged forests (Tin et al. 2018; Schulz et al. 2019). There are currently not many documentations on the soil microbiome of soils with potential BSR disease suppression, with the exception of the Blenheim soil.

The **abiotic elements** also contribute to disease suppression as the various soil pH (Fierer and Jackson 2006), carbon content (Yu et al. 2023), and clay-mineral composition (Alabouvette 1999) in the soils, have all been found to influence the pathogen (Fig. 6.3). Disease suppressive soils in Morens, Switzerland, have morainic deposits that are high in vermiculite clay, which are known to suppress the development of tobacco black root rot disease caused by *Berkeleyomyces basicola* (Almario et al. 2013, 2014). For suppressive soils in Vyklantice, Czech Republic, varying levels of N, S and Fe in the suppressive soils influenced the degree of disease suppression against the potato scab disease by *Streptomyces* spp. (Kopecky et al. 2021). In recent years, soil moisture and nutrient levels have also been found to influence disease suppressive soils as these factors are associated with the abundance of other soil microbial communities such as archaebacteria and protist communities that may have roles in nutrient recycling (Kopecky et al. 2019).

7 Suppressive Soils for BSR Disease of Oil Palm

The concept and application of disease suppressive soils is relatively novel for the oil palm industry. To date, only the **Blenheim soil** discovered in Malaysia, is deemed to be naturally supressive towards BSR disease of oil palms. The Blenheim soil (Typic Quartzipsamment) is distinctive, owing to its high deposits of calcareous shells in the soil (Goh et al. 2020, 2022). The Blenheim soil is part of the Malaysian coastal soils that have a long-established history of BSR disease. The Blenheim soil experiences the tropical climate with an annual rainfall estimated at 1700–2300 mm (Goh et al. 2020). Found in an oil palm plantation in the state of Perak, Malaysia, it has high calcareous shell deposits, is light textured, comprise primarily of coarse sand to loamy coarse sand texture, and has a relatively high pH. A census performed in 2018 revealed that BSR disease incidence were less than 2% compared to a non-suppressive soil (Bernam soil) which has reported more than 30% disease incidence (Goh et al. 2020).

7.1 Microbiota of Oil Palm Suppressive Soil

In early 2020, Goh et al. (2020) deciphered the microbiota present in the Blenheim soil using Illumina 16S rRNA and 18S rRNA amplicon sequencing. A comparison was also made with the microbiota of Bernam soil (Typic Endoaquept—non-acid sulfate) (considered as conducive soil with 33.1% disease incidence), to analyse the differences in microbial composition of both suppressive and conducive soils. Both soils were planted with Tenera (Dura Pisifera) oil palms in 1998, adopting the triangle planting system (equidistant of 8.80 × 8.80 × 8.80 m). It was revealed that the Blenheim soil has higher diversity of **prokaryotes** and eukaryotes, especially with a higher abundance of bacterial taxa that are known to have roles in suppressing disease development. This include the Calditrichaeota, GAL15, JdFR-76 (Calditrichaeota), Rokubacteria, Rubrobacter (Actinobacteria), AKYG587 (Planctomycetes), Omnitrophicaeota, and Zixibacteria. Rubrobacter (Phylum: Actinobacteria) and AKYG587 (Phylum: Planctomycetes), are also known to be prevalent in soils suppressive towards Fusarium wilt as they have strong fungistatic effect towards pathogenic *Fusarium graminearum* (Wu et al. 2008; Lee et al. 2017a, b; Siegel-Hertz et al. 2018). AKYG587 on the other hand, is potentially a helper bacterium or a symbiont, as they often have higher relative abundance when biocontrol agents such as *Pseudomonas* or *Bacillus* is inoculated (He et al. 2019; Zhao et al. 2019). For the disease conducive Bernam soil, a different soil microbiota was observed thereby establishing that the soil microbial community have an influence on Ganoderma disease incidence. It was revealed that Chloroflexi was more prevalent and these phyla have also been associated with disease conducive soils such as in soils where tobacco disease was high (Niu et al. 2016; Gao et al. 2019). A reduction in the relative abundance of Chloroflexi is observed when biocontrol agents are introduced (He et al. 2019; Zhao et al. 2019). Aside from Chloroflexi, Acidothermus (Phylum: Actinobacteria) was also detected from the Bernam soil. The role of Acidothermus is not distinctively described but their higher abundance in disease conducive soils have been noted (Svenningsen et al. 2018; Gao et al. 2019). In Goh et al. (2022), the prevalence of Acidothermus in Bernam soil may have contributed to the high BSR incidences in oil palms.

Likewise, the **eukaryotic communities** were also different in both the Blenheim and Bernam soils. In Blenheim soil, disease suppression may be associated with the higher abundance of several of these eukrayotic taxa which include Cercomonas (Phylum: Cercozoa; order: Cercomonadida) and Calcarisporiella (Phylum: Zygomycota; order: Calcarisporiellales). Cercomonas is reportedly able to feed on some bacterial groups while allowing the proliferation of bacterial antagonists such as Pseudomonads (Rønn et al. 2002; Pedersen et al. 2009, 2010, 2011). Cercomonas can therefore grow together with antagonistic bacteria while feeding on other bacteria. It is postulated that the feeding on other bacterial groups may reduce the competitiveness towards Pseudomonads, allowing the Pseudomonads to proliferate, thereby helping to suppress disease. By contrast, the disease conducive Bernam soil have higher abundance of Archaeplastida, Opisthokonta and Uronema (Phylum: Ciliophora; order: Conthreep). Information on how these various taxa influence disease development is however, not known. Nevertheless, it is suggestive that the various eukaryotic communities do have an influence on disease suppressiveness towards the pathogenic *G. boninense*.

From the microbiome analysis, it is evident that high microbial diversity is known to benefit disease suppression as lower incidence of soil-borne diseases are often observed in these suppressive soils (Chung 2011; Khunaw et al. 2017; Bokulich et al. 2018). High diversity translates to a larger pool of diverse genes that may confer beneficial properties. This was reported in the disease suppressive Blenheim soil where several rare phyla were revealed. This includes phyla such as Calditrichaeota, Zixibacteria, GAL15, Omnitrophicaeota, and Rokubacteria, described as candidate phyla radiation (CPR) (Lin et al. 2012; Castelle et al. 2013; Marshall et al. 2017; Becraft et al. 2017; Goh et al. 2019). These phyla are not well explored and their discovery in Blenheim soil is the first report. Nevertheless, some of the phyla such as Zixibacteria and Rokubacteria, are known to have highly versatile metabolic capabilities, such as in iron reducing-oxidation pathways. They are also shown through genomic evidence, to harbour genes encoding for compounds or secondary metabolites or peptides that have antimicrobial properties (Castelle et al. 2013; Crits-Christoph et al. 2018). Microbiomes with antibiotic production potential are highly likely to confer disease suppression, and this is typically linked to the phyla Alphaproteobacteria, Cyanobacteria, Firmicutes, and the genera *Pseudomonas* and *Bacillus* (Goh et al. 2022). The phylum Calditrichaeota are also known to produce exogenous peptidases that can contribute to disease suppression (Marshall et al. 2017). Therefore, the detection of these various phyla in the soil can indicate the potential of the soil for disease suppression.

7.2 *Abiotic Characteristics of Oil Palm Suppressive Soil*

The **abiotic components** of a typical suppressive soil are expected to have similar influence for soils that are suppressive to oil palm disease. As with most suppressive soils, the soil physicochemical properties (e.g., texture, pH, water potential), their mineralogy and the availability of certain nutrients and cations (e.g., aluminium, iron, and manganese), as well as composition of organic matter, have all demonstrated a certain degree of influence on the ability of the soil in suppressing disease (Janvier et al. 2007; Legrand et al. 2019; Thangavelu et al. 2020; Kopecky et al. 2021). For typical oil palm plantations, the soil profile includes heavy soils with high water retention capacity and poor drainage efficiency (Chong et al. 2017; Paterson 2019). This unfortunately, is also closely associated to low soil pH conditions. These soil profiles are prevalent in peat soils and acid sulphate soils that are found along the coastal areas in Malaysia and Indonesia, which are often converted for oil palm planting. The incidence of BSR disease is, therefore, high in most of these plantations (Corley and Tinker 2016; Khunaw and Radziah 2020; Siddiqui et al. 2021; Wong et al. 2021).

The distinctive nature of Blenheim soil in suppressing BSR incidence attracted a great interest in deciphering their soil properties. It has been hypothesized that there may be a degree of uniqueness to the soil properties of the Blenheim soil, that may very well confer several benefits that allows for disease suppression. It was revealed that Blenheim soil have high soil pH as well as a higher content of Ca. This was favourable for the growth of bacteria such as Rokubacteria and Zixibacteria (of the Candidate phyla radiation, CPR) as their abundance is noted to correspond with the higher soil pH and Ca content in the Blenheim soil. Consequently, the favoured growth of CPR bacteria may have contributed to disease suppressiveness as they are known to produce active metabolites that are capable of inhibiting pathogens. Additionally, the abundance of several other beneficial phyla (GAL15, Omnitrophicaeota, Calditrichaeota) in Blenheim soil, has been linked to the calcareous parent material and the higher soil pH (~pH 7) (Goh et al. 2020).

Comparatively, the abiotic components of the Blenheim soil showed a different profile to the non-suppressive soil (i.e. Bernam soil), further explaining the greater disease suppression observed in the Blenheim soil. In Goh et al. (2020), the principal component analysis (PCA) revealed that soil pH, as well as soil nutrient content (N, Ca, Cu, TP), and soil texture (coarse sand) were significantly varied from the same parameters measured for the Bernam soil (Goh et al. 2020). The Blenheim suppressive soil is observed to have a light textural class, coarse sandy nature, higher soil pH (pH 7–8) as well as high calcium content compared to the Bernam soil. The coarse sandy composition of Blenheim soil renders the soil with low water retention capability. This prevents water-logging conditions, reducing disease incidence. The high calcium content is derived from the calcareous shell deposits that constitute the origin of the Blenheim soil (Goh et al. 2020). Shells from Crustaceans and molluscs often comprise of 5–88% calcium carbonate, 5–70% chitin, and 6–46% protein (Abdou et al. 2008; Alabaraoye et al. 2018). As such, the calcareous shell deposits in the Blenheim soil provided 44% of the total calcium found in the soil (Goh et al. 2020). The high calcareous shell deposits in Blenheim soil are postulated to render many benefits. It has been reported that crustacean shells have high chitin and the availability of chitin biopolymers in soil have been known to improve plant growth and inducing plant defence response towards disease (Ilangumaran et al. 2017; Sharp 2013). The high chitin rich environment may also likely support the growth of microbes with the ability to produce exogenous chitinases and proteases, which are also key enzymes for defence against plant fungal diseases (Punja and Zhang 1993; Ilangumaran et al. 2017; Inderbitzin et al. 2018). It is highly likely that the presence of these calcareous shell deposits in Blenheim soil may also contribute to disease suppressiveness towards BSR incidence.

The near-neutral pH of Blenheim soil is also highly desirable. These conditions are ideal for microbial communities, often leading to higher diversity and richness, and the increase of microbial biomass (Rousk et al. 2009, 2010). Higher soil pH (pH 8) also harbours greater relative abundance of bacterial diversity compared to acidic soils (pH 4) (Rousk et al. 2010). Similarly, bacterial diversity is greater in soils with calcareous parent materials such as from the Blenheim soil. On the contrary, lower soil pH favours higher disease incidence. Acidic soils often have aluminium ions that encourages aluminium toxicity. As a result, microbial activities are limited in acidic soils, affecting the functions of beneficial bacteria such as rhizobacteria, siderophore-producing Pseudomonads, and biocontrol *Bacillus* spp. (Ownley et al. 2003; Li et al. 2017). Conversely, acidic soils favour the growth of *Ganoderma boninense*, resulting in higher BSR disease incidence in oil palms planted in acidic soils compared to soils with near neutral pH (Nawawi and Ho 1990; Rahman and Othman 2020). Other soil factors such as salinity and sodium chloride concentration have also been proposed as likely factors contributing to soil disease suppression. It has been reported that soil salinity and the level of sodium chloride in the soil can influence the growth of *Cercomonas* and *Uronema*. *Cercomonas* species are inhibited by salinity levels of 5–10 parts per thousands (ppt) (Ekelund 2002) while *Uronema* species prefer a salinity range of 17–43 ppt salinity (Hamilton and Preslan 1969). The distribution of both the species in the soil influences the dynamics of the microbial communities as they interact with other microbes especially beneficial microbes, and subsequently, the degree of disease suppression.

8 Artificially Induced Suppressive Soils

Naturally occurring suppressive soils for disease control are limited or mostly undiscovered. As such, there have been attempts to turn some of the conducive soils into suppressive soils by artificially inducing disease suppressiveness in the conducive soils. To turn conducive soils into disease suppressive soils, both biotic and abiotic components are often manipulated (Cha et al. 2016; Mousa and Raizada 2016) (Fig. 6.3). The **biotic components** can be improved by introducing microbial isolates that are beneficial and able to suppress disease development. Introduction of beneficial microbes such as species of *Streptomyces, Bacillus, Pseudomonas, Penicillium* and *Trichoderma*, have been shown to inhibit the growth of various soil-borne pathogens and can be useful to confer disease suppression. The core strategy in developing a successful artificially induced suppressive soil is to decipher beneficial microbes from soils with potential for disease suppression (Weller et al. 2002). These microbes are then intentionally introduced into conducive soils to enhance the rhizosphere or bulk soil microbiome to contribute to disease suppression. Typically, these beneficial microbes demonstrate attributes to produce inhibitory compounds, compete or hyperparasitize the pathogens. They also stimulate plant growth and vigour, and elicit induce host resistance as mechanisms to suppress disease incidence (Kielak et al. 2016; Mousa and Raizada 2016; Schlatter et al. 2017; Liu et al. 2020). Another critical point in developing effective disease suppressive soils is to maintain a balance between the microbial diversity and microbial biomass (Weller et al. 2002; Schlatter et al. 2017). This balance ensures that the pathogens remain at relatively lower counts (Benizri et al. 2005), thereby preventing their subsequent colonization and establishment in the soil (Weller et al. 2002; Hadar and Papadopoulou 2012; Schlatter et al. 2017).

For **abiotic components**, the soil pH, soil organic matter (SOM), and clay composition can be augmented to influence the population, diversity and microbial activities of the soil microbiota (Sullivan 2001; Peters et al. 2003) (Fig. 6.3). The development of artificially induced suppressive soils benefits in controlling several diseases that are relatively difficult to control with other means. For example, some diseases caused by fungi, Oomycetes, or root rot pathogens, are difficult to manage with the use of fungicides and resistant cultivars, but benefit from disease suppressive soils (Persson and Olsson 2000; Jambhulkar et al. 2015). Additionally, several strategies can be employed to ensure disease suppressiveness is sustained in the suppressive soil. This includes manipulating soil microbiota by incorporating amendments such as liming (Murakami et al. 2002) or composts (Hadar and Papadopoulou 2012), or by the introduction of biological control agents (Castãno et al. 2013; de Corato et al. 2018). The addition of soil organic amendments is usually effective in helping the microbial community to enhance their functionality and gradual disease suppression. Introduction of beneficial microbiota in the form of biological control agents (BCAs) as well as the augmentation using organic amendments (OAs) and composts, appears to be promising strategies as well to enhance soil disease suppression (Trillas et al. 2006; de Corato et al. 2018).

9 Developing Suppressive Soils for Oil Palm

The strategies to develop artificially induced suppressive soils for oil palm diseases is very much similar to developing disease suppressive soils for other diseases. The primary approaches are through the manipulation of soil pH and the introduction of beneficial microbes especially biocontrol agents. Soil pH that is increased with the application of soil amendments such as calcium carbonate (He et al. 2014), is known to have a positive impact on the growth of bacteria responsible for disease suppression such as *Bacillus* (Liu et al. 2019), *Paenibacillus* (Haggag and Timmusk 2008), *Arthrobacter* (Xiao et al. 2018), *Flavobacterium* (Kwak et al. 2018), *Pseudomonas* (Raaijmakers and Weller 1998), and *Streptomyces* (Cha et al. 2016). Increasing soil pH to pH 6–6.5 (Zhang et al. 2022) has also been observed as helpful in promoting disease suppression. Soil pH is known to influence microbial communities via the influence of pH on bioavailability of key nutrients (e.g., carbon, nitrogen, phosphorous, sulfur) and cations (e.g., aluminium, calcium, magnesium) (Lemanceau et al. 2015). These nutrients and cations are essential to support the growth of beneficial microbes. Soil pH is therefore a critical factor for the success of disease suppression.

Disease suppressiveness towards BSR disease can also be developed by the introduction of biological control agents. Similar to other diseases, the introduction of biocontrol agents seemed to benefit in most cases, and the application of more than a single microbial agent is often able to generate better efficacy of disease suppression. In artificially induced suppressive soils, a combination of bacterial-bacterial, fungal-fungal, or bacterial-fungal biocontrol agent can often render satisfactory outcome (Singh 1991; Chung 2011; Norman et al. 2019). The mixed cultures interact with one another to support growth and to adapt better to environmental conditions (Naher et al. 2013; Muniroh et al. 2019). Mixed cultures of various groups of beneficial microbes which includes plant growth promoting bacteria, arbuscular mycorrhizal fungi, mycoparasitic *Trichoderma* spp., and antibiotic-producing *Penicillium* spp. are therefore desirable to produce soils with disease suppression (Sharma et al. 1992; Idris et al. 2004a; Chaparro et al. 2012; Tian et al. 2020). In a separate study, some degree of disease suppression and growth enhancement was achieved with the application of selected *Pseudomonas aeruginosa* (UPMP3) with Arbuscular Mycorrhizal Fungi (AMF) (*Glomus intraradices* UT126, *G. clarum* BR152B) (Sundram 2010; Sundram et al. 2015). The combination of these microbial isolates reduced BSR disease incidence in oil palm to 50–80%, and improved root mass by 35% (Sundram et al. 2015). It is evident that artificially induced suppressive soils for control of BSR disease can be developed using beneficial microbial isolates and manipulation of soil characteristics, although with varying degree of success.

10 Suppressive Soils for Oil Palm Disease Management-Challenges and Future Direction

The adoption of suppressive soil for disease management of oil palm, particularly for BSR disease, has several **challenges**. To achieve widespread adoption of suppressive soil as a control strategy, it would firstly require several of the current limitations to be overcome. These limitations include the complex microbial interactions, the ability to upscale this technology, the unpredictable susceptibility of this approach to climate change conditions, and the current research gap in understanding suppressive soils. As suppressive soils are predominantly regulated by the microbiome, their complex interaction with one another and with the pathogen can influence the efficacy of disease suppression. The interaction of the microbiome therefore requires further studies as the interactions could be synergistic or antagonistic, thereby influencing the outcome of the disease suppression. The other challenge in adopting suppressive soil as a means of control is the up-scaling of this technology. Up-scaling and the transfer of this technology would require cost-effective measures and amenability for transfer to various commercial plantations. The use of suppressive soil is also influenced by climate change conditions. Rainfall, temperature, land use, are all known to affect soil properties and microbial activities. With climate change, these factors have a greater magnitude of severity, impacting the efficacy of soil disease suppression. Additionally, the adoption of suppressive soil will also require careful planning of soil management practices. Land use and cultural practices, in the form of anthropogenic factors, can also influence soil as these activities impact nutrient cycling and soil physical properties (Anothai and Chairin 2022). There is also a gap in the current research of suppressive soils. Although suppressive soil is an attractive means of control strategy, there are limited suppressive soils known and even lesser information on their characteristics and use. There may be new soils which potentially have disease suppression effect towards other pathogens of oil palm, but the research in discovering these suppressive soils is progressing slowly and lack intensity. For example, Rusli et al. (2016) have evaluated several Malaysian soils that have shown potential for disease suppression towards Fusarium wilt of oil palm (by pathogenic *Fusarium oxysporum* f. sp. *elaeidis*). However, there are no further reports on the evaluated suppressive soil, their application and efficacy thereafter.

Future research on areas related to soil biology is warranted to obtain more information on the full potential of suppressive soils especially for oil palm cultivation. Disease suppressive soil is a natural soil characteristic, therefore, this approach strongly advocates the sustainable and environmentally friendly approach to crop management (de Corato 2020a). There are several strategies that can be implemented to enhance the efficacy of soil disease suppressiveness. Practices such as incorporating compost can be beneficial as composts are known to influence the microbial communities, especially beneficial microbes, thereby rendering soils suppressive towards the pathogen (Mehta et al. 2014; Chilosi et al. 2017; de Corato 2020a). De Corato (2020b) have also reported that species richness and diversity is enhanced through application of composts (e.g., agro-wastes), which successfully suppresses disease development. Some soils benefit from application of biofertilizers as well. Biofertilizers elicit change in soil microbiome that can lead to a strong reduction of disease incidence. In some soil management practices, crop rotation and intercropping are considered beneficial for suppressive soil. This cropping practice is much more efficient than monocropping as soil chemical properties and microbiome are affected by long-term monocropping system which can cause economic losses (Zhao et al. 2018). Monocropping can trigger soil acidification as organic acids accumulate in rhizosphere and soil, which consequently impact the abundance, diversity and structure of the microbial community (Bai et al. 2019). The practice of crop rotation or intercropping can therefore aid in mitigating impacts to soil and microbial properties. With changes to vegetation, the association between plant-microbe-soil also changes, as the microbiome shifts are dependent on microbial consortia and soil amendments applied. Intercropping therefore helps in altering the soil microbiome as some specific plants produce exudates that are inhibitory towards pathogens (Li et al. 2013).

11 Conclusion

As with many other agricultural crops, suppressive soils are equally important for oil palm management. Suppressive soils are sustainable and innovative and is ideal for the environmentally friendly approach for management of oil palm cultivation. Soil microbiomes are critical to the effectiveness of disease suppressive soils. Suppressive soil is highly favoured as an eco-friendly approach. Suppressive soil can help oil palm growers increase yield while reducing environmental impact and ensuring long-term viability. As the oil palm industry continues to grow, the demand for sustainable practices increases. By adopting environmentally friendly approaches such as the use of suppressive soil, this industry can benefit from suppressive soil technology offering a positive outcome for farmers, consumers, and the planet.

Acknowledgement The author expresses gratitude to Monash University Malaysia for the facilities and financial support for the author's research endeavor on suppressive soil. And to Advanced Agriecological Research Sdn. Bhd. for the research grant on deciphering soil microbiome of the Blenhiem suppressive soil.

References

Abdou ES, Nagy KSA, Elsabee MZ (2008) Extraction and characterization of chitin and chitosan from local sources. Bioresour Technol 99(5):1359–1367

Alabaraoye E, Achilonu M, Hester R (2018) Biopolymer (chitin) from various marine seashell wastes: isolation and characterization. J Polym Environ 26(6):2207–2218

Alabouvette C (1986) Fusarium wilt suppressive soils from the Chateaurenard region: review of a 10-year study. Agronomie 6:273–284. https://doi.org/10.1051/agro:19860307

Alabouvette C (1999) Fusarium wilt suppressive soils: an example of disease suppressive soils. Australas Plant Pathol 28:57–64

Almario J, Kyselkova M, Kopecky J, Sagova-Mareckova M, Muller D, Grundmann GL, Moenne-Loccoz Y (2013) Assessment of the relationship between geologic origin of soil, rhizobacterial community composition and soil receptivity to tobacco black root rot in Savoie region (France). Plant Soil 371:397–408

Almario J, Muller D, Defago G, Moenne-Loccoz Y (2014) Rhizosphere ecology and phytoprotection in soils naturally suppressive to Thielaviopsis black root rot of tobacco. Environ Microbiol 16:1949–1960

Andrade OA, Mathre DE, Sands DC (1994) Suppression of *Gaeumannomyces graminis* var. *tritici* in Montana soils and its transferability between soils. Soil Biol Biochem 26:397–402. https://doi.org/10.1016/0038-0717(94)90289-5
Anothai J, Chairin T (2020) Soil physicochemical properties closely associated with fungal enzymes and plant defense enzymes in Ganoderma-infected oil palm orchards. Plant Soil 456: 99–112. https://doi.org/10.1007/s11104-020-04705-y
Anothai J, Chairin T (2022) Analysis of rhizobacterial community associated with the occurrence of Ganoderma basal stem rot disease in oil palm by Illumina next-generation sequencing. Arch Microbiol 204:31. https://doi.org/10.1007/s00203-021-02670-3
Bai Y, Wang G, Cheng Y, Shi P, Yang C, Yang H, Xu Z (2019) Soil acidification in continuously cropped tobacco alters bacterial community structure and diversity via the accumulation of phenolic acids. Sci Rep 9:12499. https://doi.org/10.1038/s41598-019-48611-5
Baker K, Cook RJ (1974) Biological control of plant pathogens. Freeman WH Publisher, San Francisco
Barcelos E, Rios SDA, Cunha RN, Lopes R, Motoike SY, Babiychuk E, Skirycz A, Kushnir S (2015) Oil palm natural diversity and the potential for yield improvement. Front Plant Sci 6:190
Becraft ED, Woyke T, Jarett J, Ivanova N, Godoy-Vitorino F, Poulton N, Brown JM, Brown J, Lau MCY, Onstott T et al (2017) Rokubacteria: genomic giants among the uncultured bacterial phyla. Front Microbiol 8:2264
Benizri E, Piutti S, Verger S, Pages L, Vercambre G, Poessel J, Michelot P (2005) Replant diseases: bacterial community structure and diversity in peach rhizosphere as determined by metabolic and genetic fingerprinting. Soil Biol Biochem 37:1738–1746. https://doi.org/10.1016/j.soilbio.2005.02.009
Bentivoglio D, Finco A, Bucci G (2018) Factors affecting the Indonesian palm oil market in food and fuel industry: evidence from a time series analysis. Int J Energy Econ Policy 8(5):49–57
Bokulich NA, Kaehler BD, Rideout JR, Dillon M, Bolyen E, Knight R, Huttley GA, Gregory Caporaso J (2018) Optimizing taxonomic classification of marker-gene amplicon sequences with QIIME 20s q2-feature-classifier plugin. Microbiome 6:90
Carlos C, Ricardo E (2004) Red ring and others disease of the oil palm in Central and South America. Proceedings of the international conference on pests and diseases of importance to the oil palm industry
Castãno R, Borrero C, Trillas MI, Avilés M (2013) Selection of biological control agents against tomato Fusarium wilt and evaluation in greenhouse conditions of two selected agents in three growing media. BioControl 58:105–116
Castelle CJ, Hug LA, Wrighton KC, Thomas BC, Williams KH, Wu D, Tringe SG, Singer SW, Eisen JA, Banfield JF (2013) Extraordinary phylogenetic diversity and metabolic versatility in aquifer sediment. Nat Commun 4:2120
Cha JY, Han S, Hong HJ, Cho H, Kim D, Kwon Y, Kwon SK, Crüsemann M, Bok LY, Kim JF, Giaever G, Nislow C, Moore BS, Thomashow LS, Weller DM, Kwak YS (2016) Microbial and biochemical basis of a Fusarium wilt-suppressive soil. ISME J 10(1):119–129
Chandrashekara C, Kumar R, Bhatt JC, Chandrashekara KN (2012) Supressive soils in plant disease management. In: Singh VK, Singh Y, Singh A (eds) Eco-friendly innovative approaches in plant disease management. International Book Distributors, India, pp 241–256. https://doi.org/10.13140/2.1.5173.7608
Chaparro JM, Sheflin AM, Manter DK, Vivanco JM (2012) Manipulating the soil microbiome to increase soil health and plant fertility. Biol Fertil Soils 48:489–499
Chilosi G, Aleandri MP, Bruni N, Tomassini A, Torresi V, Muganu M, Paolocci M, Vettraino A, Vannini A (2017) Assessment of suitability and suppressiveness of on-farm green compost as a substitute of peat in the production of lavender plants. Biocontrol Sci Tech 27(4):539–555
Chong KP, Dayou J, Alexander A (2017) Pathogenic nature of *Ganoderma boninense* and basal stem rot disease. In: Chong KP, Dayou J, Alexander A (eds) Detection and control of *Ganoderma boninense* in oil palm crop. Springer International Publishing, Cham, pp 5–12

Chung G (1990) Preliminary results on trunk injection of fungicides against Ganoderma basal stem rot in oil palm. Proceedings of Ganoderma workshop, Bangi, Selangor, Malaysia, 11 September 1990, pp 81–97

Chung GF (2011) Management of *Ganoderma* diseases in oil palm plantations. Planter 87:325–339

Cook RJ (2014) Plant health management: pathogen suppressive soils. In: Alfen V, Neal K (eds) Encyclopedia of agriculture and food systems. Academic, pp 441–455. https://doi.org/10.1016/B978-0-444-52512-3.00182-0

Corley RHV (2009) How much palm oil do we need? Environ Sci Pol 12:134–139. https://doi.org/10.1016/j.envsci.2008.10.011

Corley RHV, Tinker PB (2003) The Oil Palm, 4th ed.; Blackwell Science: Oxford, UK, pp. 25–26

Corley RHV, Tinker PB (2016) The oil palm, 5th edn. Blackwell Publishing, Oxford

Crits-Christoph A, Diamond S, Butterfield CN, Thomas BC, Banfield JF (2018) Novel soil bacteria possess diverse genes for secondary metabolite biosynthesis. Nature 558:440–444

Darmono T (2000) Ganoderma in oil palm in Indonesia: current status and prospective use of antibodies for the detection of infection. In: Flood J, Bridge PD, Holderness M (eds) Ganoderma diseases of perennial crops. Wallingford, UK: CABI Publishing, pp 249–266. https://doi.org/10.1079/9780851993881.0249

De Corato U (2020a) Disease-suppressive compost enhances natural soil suppressiveness against soil-borne plant pathogens: a critical review. Rhizosphere 13:100192. https://doi.org/10.1016/j.rhisph.2020.100192

De Corato U (2020b) Soil microbiota manipulation and its role in suppressing soil-borne plant pathogens in organic farming systems under the light of microbiome-assisted strategies. Chem Biol Technol Agric 7:17. https://doi.org/10.1186/s40538-020-00183-7

De Corato U, Salimbeni R, De Pretis A, Patruno L, Avella N, Lacolla G, Cucci G (2018) Microbiota from 'next-generation green compost' improves suppressiveness of composted municipal-solid-waste to soil-borne plant pathogens. Biol Control 124:1–17. https://doi.org/10.1016/j.biocontrol.2018.05.020

Dislich C, Keyel AC, Salecker J, Kisel Y, Meyer KM, Auliya M, Barnes AD, Corre MD, Darras K, Faust H (2017) A review of the ecosystem functions in oil palm plantations, using forests as a reference system. Biol Rev 92:1539–1569

Ekelund F (2002) Tolerance of soil flagellates to increased NaCl levels. J Eukaryot Microbiol 49:324–328

Expósito RG, de Bruijn I, Postma J, Raaijmakers JM (2017) Current insights into the role of rhizosphere bacteria in disease suppressive soils. Front Microbiol 8:1–12. https://doi.org/10.3389/fmicb.2017.02529

Fathana H (2018) Palm oil politics in Malaysia and Indonesia: competition or collaboration? JATI J SouthEast Asian Stud 23:47–64

Fierer N, Jackson RB (2006) The diversity and biogeography of soil bacterial communities. PNAS 103(3):626–631. https://doi.org/10.1073/pnas.0507535103

Gao L, Wang R, Gao J, Li F, Huang G, Huo G, Liu Z, Tang W, Shen G (2019) Analysis of the structure of bacterial and fungal communities in disease suppressive and disease conducive tobacco-planting soils in China. Soil Res 58:35–40

Goh KM, Shahar S, Chan KG, Chong CS, Amran SI, Sani MH, Zakaria II, Kahar UM (2019) Current status and potential applications of underexplored prokaryotes. Microorganisms 7:468

Goh YK, Md Zoqratt MZH, Goh YK, Ayub Q, Ting ASY (2020) Determining soil microbial communities and their influence on Ganoderma disease incidences in oil palm (*Elaeis guineensis*) via high-throughput sequencing. Biology 9(12):424. https://doi.org/10.3390/biology9120424

Goh YK, MdZoqratt MZH, Goh YK, Ayub Q, Ting ASY (2022) Discovering naturally-occurring microbiota in disease suppressive soil: potential role of biological elements in suppressing *Ganoderma boninense*. Biol Control 165:104787. https://doi.org/10.1016/j.biocontrol.2021.104787

Gomez ER, de Bruijn I, Postma J, Raaijmakers JM (2017) Current insights into the role of rhizosphere bacteria in disease suppressive soil. Front Microbiol 8:2529

Guerra CA, Heintz-Buschart A, Sikorski J, Chatzinotas A, Guerrero-Ramírez N, Cesarz S, Beaumelle L, Rillig MC, Maestre FT, Delgado-Baquerizo M, Buscot F (2020) Blind spots in global soil biodiversity and ecosystem function research. Nat Commun 11(1):1–13. https://doi.org/10.1038/s41467-020-17688-2

Hadar Y, Papadopoulou KK (2012) Suppressive composts: microbial ecology links between abiotic environments and healthy plants. Annu Rev Phytopathol 50(1):133–153

Haggag WM, Timmusk S (2008) Colonization of peanut roots by biofilm forming *Paenibacillus polymyxa* initiates biocontrol against crown rot disease. J Appl Microbiol 104:961–969. https://doi.org/10.1111/j.1365-2672.2007.03611.x

Hamilton RD, Preslan JE (1969) Cultural characteristics of a pelagic marine hymenostome ciliate, *Uronema* sp. J Exp Mar Biol Ecol 4:90–99

He K, Yang SY, Li H, Wang H, Li ZL (2014) Effects of calcium carbonate on the survival of *Ralstonia solanacearum* in soil and control of tobacco bacterial wilt. Eur J Plant Pathol 140:665–675. https://doi.org/10.1007/s10658-014-0496-4

He Y, Wu Z, Wang W, Liu X, Ye BC (2019) Bacterial community and phosphorus species changes in pepper rhizosphere soils after *Pseudomonas putida* Rs-198 inoculation. Rhizosphere 11:100164

Hushiarian R, Yusof NA, Dutse SW (2013) Detection and control of *Ganoderma boninense*: strategies and perspectives. Springerplus 2:555

Idris A, Kushairi A, Ismail S, Ariffin D (2004a) Selection for partial resistance in oil palm progenies to *Ganoderma* basal stem rot. J Oil Palm Res 16:12–18

Idris A, Ahmad H, Ariffin D, Ismail S (2004b) Prolonging the productive life of Ganoderma-infected palms with hexaconazole. MPOB TS Info Ser:2004

Idris A, Rahman RA, Kushairi A (2010) Hexaconale as a preventive treatment for managing Ganoderma in oil palm. MPOB TS Info Ser 75

Idris AS, Mior MHAZ, Maizatul SM, Kushairi A (2011) Survey on status of Ganoderma disease of oil palm. In: Proceedings of the PIPOC 2017 international palm oil congress: agriculture conference, pp 235–238

Ilangumaran G, Stratton G, Ravichandran S, Shukla PS, Potin P, Asiedu S, Prithiviraj B (2017) Microbial degradation of lobster shells to extract chitin derivatives for plant disease management. Front Microbiol 8:781

Inderbitzin P, Ward J, Barbella A, Solares N, Izyumin D, Burman P, Chellemi DO, Subbarao KV (2018) Soil microbiomes associated with Verticillium wilt-suppressive broccoli and chitin amendments are enriched with potential biocontrol agents. Phytopathology 108(1):31–43

Irma A, Meryandini A, Rupaedah B (2018) Biofungicide producing bacteria: an in vitro inhibitor of *Ganoderma boninense*. HAYATI J Biosci 25:151–159

Jambhulkar PP, Sharma M, Lakshman D, Sharma P (2015) Natural mechanisms of soil suppressiveness against diseases caused by *Fusarium, Rhizoctonia, Pythium*, and *Phytophthora*. In: Meghvansi MK, Varma A (eds) Organic amendments and soil suppressiveness in plant disease management. Springer, New York, pp 95–124

Janvier C, Villeneuve F, Alabouvette C, Edel-Hermann V, Mateille T, Steinberg C (2007) Soil health through soil disease suppression: which strategy from descriptors to indicators? Soil Biol Biochem 39:1–23

Kalam MA, Masjuki HH (2002) Biodiesel from palm oil-an analysis of its properties and potential. Biomass Bioenergy 23:471–479

Khunaw AR, Radziah O (2020) Influence of pH levels on disease development in oil palm seedling roots infected with *Ganoderma boninense*. Rhizosphere 13:100181

Khunaw A, Othman R, Ali N, Musa M, Rabara F (2017) Role of microbial community in suppressing development of *Ganoderma* in oil palm seedlings. Int J Plant Soil Sci 19(1):1–14

Kielak AM, Cipriano MAP, Kuramae EE (2016) Acidobacteria strains from subdivision 1 act as plant growth-promoting bacteria. Arch Microbiol 198(10):987–993

Koh LP (2007) Potential habitat and biodiversity losses from intensified biodiesel feedstock production. Conserv Biol 21:1373–1375. https://doi.org/10.1111/j.1523-1739.2007.00771.x

Kopecky J, Samkova Z, Sarikhani E, Kyselková M, Omelka M, Kristufek V et al (2019) Bacterial, archaeal and micro-eukaryotic communities characterize a disease-suppressive or conducive soil and a cultivar resistant or susceptible to common scab. Sci Rep 9(1):1–14. https://doi.org/10.1038/s41598-019-51570-6

Kopecky J, Rapoport D, Sarikhani E, Stovicek A, Patrmanova T, Sagova-Mareckova M (2021) Micronutrients and soil microorganisms in the suppression of potato common scab. Agronomy 11(2):383. https://doi.org/10.3390/agronomy11020383

Kushairi A, Loh SK, Azman I, Hishamuddin E, Ong-Abdullah M, Izuddin Z, Razmah G, Sundram S, Parveez GKA (2018) Oil palm economic performance in Malaysia and R&D progress in 2017. J Oil Palm Res 30:163–195

Kushairi A, Ong-Abdullah M, Nambiappan B, Hishamuddin E, Bidin MNIZ, Ghazalia R, Subramaniam V, Sundram S, Parveez GKA (2019) Oil palm economic performance in Malaysia and R&D progress in 2018. J Oil Palm Res 31:165–194

Kwak MJ, Kong HG, Choi K, Kwon SK, Song JY, Lee J, Lee PA, Choi SY, Seo M, Lee HJ, Jung EJ, Park H, Roy N, Kim H, Lee MM, Rubin EM, Lee SW, Kim JF (2018) Rhizosphere microbiome structure alters to enable wilt resistance in tomato. Nat Biotechnol 36:1100–1109. https://doi.org/10.1038/nbt.4232

Kyselkova M, Kopecký J, Frapolli M, Défago G, Sagova-Mareckova M, Grundmann GL, Moenne-Loccoz Y (2009) Comparison of rhizobacterial community composition in soil suppressive or conducive to tobacco black root rot disease. ISME J 3(10):1127–1138

Lee CG, Iida T, Inoue Y, Muramoto Y, Watanabe H, Nakaho K, Ohkuma M (2017a) Prokaryotic communities at different depths between soils with and without tomato bacterial wilt but pathogen-present in a single greenhouse. Microbes Environ 32:118–124

Lee CG, Iida T, Uwagaki Y, Otani Y, Nakaho K, Ohkuma M (2017b) Comparison of prokaryotic and eukaryotic communities in soil samples with and without tomato bacterial wilt collected from different fields. Microbes Environ 32:376–385

Legrand F, Chen W, Cobo-Díaz Jose F, Picot A, Floch GL (2019) Co-occurrence analysis reveal that biotic and abiotic factors influence soil fungistasis against *Fusarium graminearum*. FEMS Microbiol Ecol 95(5). https://doi.org/10.1093/femsec/fiz056

Lemanceau P, Maron PA, Mazurier S, Mougel C, Pivato B, Plassart P, Ranjard L, Revellin C, Tardy V, Wipf D (2015) Understanding and managing soil biodiversity: a major challenge in agroecology. Agron Sustain Dev 35:67–81. https://doi.org/10.1007/s13593-014-0247-0

Li S, Liu Y, Wang J, Yang L, Zhang S, Xu C, Ding W (2017) Soil acidification aggravates the occurrence of bacterial wilt in South China. Front Microbiol 8:703

Liew WL, Kassim MA, Muda K, Loh SK, Affam AC (2015) Conventional methods and emerging wastewater polishing technologies for palm oil mill effluent treatment: a review. J Environ Manag 149:222–235

Lim T, Hamm R, Mohamad R (1990) Persistency and volatile behaviour of selected chemicals in treated soil against three basidiomycetous root disease pathogens. Int J Pest Manag 36:23–26

Lim T, Chung G, Ko W (1992) Basal stem rot of oil palm caused by *Ganoderma boninense*. J Plant Pathol 1:147–152

Lim PH, Gansau JA, Chong KP (2018) *Streptomyces* spp. a potential biocontrol agent against *Ganoderma boninense* of basal stem rot. J Oil Palm Res 30:265–275

Lin X, Kennedy D, Fredrickson J, Bjornstad B, Konopka A (2012) Vertical stratification of subsurface microbial community composition across geological formations at the Hanford Site. Environ Microbiol 14:414–425

Lisnawita HH, Tantawi AR (2016) Survey of basal stem rot disease on oil palms (*Elaeis guineensis* Jacq.) in Kebun Bukit Kijang, North Sumatera, Indonesia. IOP Conf Ser Earth Environ Sci 41:012007. https://doi.org/10.1088/1755-1315/41/1/012007

Liu D, Li K, Hu J, Wang W, Liu X, Gao Z (2019) Biocontrol and action mechanism of *Bacillus amyloliquefaciens* and *Bacillus subtilis* in soybean phytophthora blight. Int J Mol Sci 20:2908. https://doi.org/10.3390/ijms20122908

Liu T, Li J, Lim KZ, Chen H, Hu S, Yuan Z, Guo J (2020) Simultaneous removal of dissolved methane and nitrogen from synthetic mainstream anaerobic effluent. Environ Sci Technol 54(12):7629–7638

Maluin FN, Hussein MZ, Idris AS (2020) An overview of the oil palm industry: challenges and some emerging opportunities for nanotechnology development. Agronomy 10:356. https://doi.org/10.3390/agronomy10030356

Marshall IPG, Starnawski P, Cupit C, Fernández Cáceres E, Ettema TJG, Schramm A, Kjeldsen KU (2017) The novel bacterial phylum Calditrichaeota is diverse, widespread and abundant in marine sediments and has the capacity to degrade detrital proteins. Environ Microbiol Rep 9:397–403

Martinez G, Sanz JI, Torres G, Sarria G, Velez D, Zuniga F, Mestizo Y, Varon F (2018) The integrated management of bud rot disease and *Phytophthora palmivora* in oil palm. Colombian Oil Palm Research Center – Cenipalma, Colombia. In: Rival A (ed) Achieving sustainable cultivation of oil palm, vol 2. Burleigh Dodds Science Publishing, London, 22 pages

Mazzola M (2002) Mechanisms of natural soil suppressiveness to soilborne diseases. Antonie Van Leeuwenhoek 81:557–564

Mehta C, Palni U, Franke-Whittle I, Sharma A (2014) Compost: its role, mechanism and impact on reducing soil-borne plant diseases. Waste Manag 34(3):607–622

Mendes R, Kruijt M, de Bruijn I, Dekkers E, van der Voort M, Schneider JHM, Piceno YM, DeSantis TZ, Andersen GL, Bakker PAHM (2011) Deciphering the rhizosphere microbiome for disease-suppressive bacteria. Science 332:1097

Mousa WK, Raizada MN (2016) Natural disease control in cereal grains. In: Wrigley C, Corke H, Seetharaman K, Faubion J (eds) Encyclopedia of food grains. Academic, Oxford, pp 257–263

MPOB (2019) Malaysian oil palm statistics 2018, 38th edn. Malaysian Palm Oil Board, Bangi

MPOB (Malaysia Palm Oil Board) (2017) Overview of the Malaysian Oil Palm Industry 2016. http://bepi.mpob.gov.my/index.php/en/statistics/area/176-area-2016/790-oil-palm-planted-abrea-as-at-dec-2016.html

Muniroh MS, Nusaibah SA, Vadamalai G, Siddique Y (2019) Proficiency of biocontrol agents as plant growth promoters and hydrolytic enzyme producers in *Ganoderma boninense* infected oil palm seedlings. Curr Plant Biol 20:100116

Murakami H, Tsushima S, Kuroyanagi Y, Shishido Y (2002) Reduction of resting spore density of *Plasmodiophora brassicae* and clubroot disease severity by liming. Soil Sci Plant Nutr 48:685–691

Naher L, Yusuf UK, Ismail A, Tan SG, Mondal MMA (2013) Ecological status of *Ganoderma* and basal stem rot disease of oil palms (*Elaeis guineensis* Jacq.). Aust J Crop Sci 7:1723–1727

Nawawi A, Ho YW (1990) Effect of temperature and pH on growth pattern of *Ganoderma boninense* from oil palm in Peninsular Malaysia. Pertanika 13:303–307

Niu J, Rang Z, Zhang C, Chen W, Tian F, Yin H, Dai L (2016) The succession pattern of soil microbial communities and its relationship with tobacco bacterial wilt. BMC Microbiol 16:233

Norman K, Idris AS, Mohd Mazmira MM (2019) Prospects in sustainable control of oil palm pests and diseases through the enhancement of ecosystem services—the way forward. J Oil Palm Res 31:381–393

Oil Palm Planted Area (2018). http://bepi.mpob.gov.my/index.php/en/statistics/area/189-area2018/857-oil-palm-planted-area-as-at-dec-2018.html

Ownley BH, Duffy BK, Weller DM (2003) Identification and manipulation of soil properties to improve the biological control performance of phenazine-producing *Pseudomonas fluorescens*. Appl Environ Microbiol 69:3333

Paterson RRM (2019) *Ganoderma boninense* disease deduced from simulation modelling with large data sets of future Malaysian oil palm climate. Phytoparasitica 47(2):255–262

Paterson RRM (2022) Future scenarios for Fusarium wilt disease and mortality of oil palm in Nigeria, Ghana and Cameroon, extrapolated to Malaysia and Indonesia. Eur J Plant Pathol 162:105–117. https://doi.org/10.1007/s10658-021-02389-3

Pedersen AL, Nybroe O, Winding A, Ekelund F, Bjørnlund L (2009) Bacterial feeders, the nematode *Caenorhabditis elegans* and the flagellate *Cercomonas longicauda*, have different effects on outcome of competition among the *Pseudomonas* biocontrol strains CHA0 and DSS73. Microb Ecol 57:501–509

Pedersen AL, Ekelund F, Johansen A, Winding A (2010) Interaction of bacteria-feeding soil flagellates and *Pseudomonas* spp. Biol Fertil Soils 46:151–158

Pedersen AL, Winding A, Altenburger A, Ekelund F (2011) Protozoan growth rates on secondary-metabolite producing *Pseudomonas* spp. correlate with high-level protozoan taxonomy. FEMS Microbiol Lett 316:16–22

Persson L, Olsson S (2000) Abiotic characteristics of soils suppressive to *Aphanomyces* root rot. Soil Biol Biochem 32:1141–1150. https://doi.org/10.1016/S0038-0717(00)00030-4

Peters RD, Sturz AV, Carter MR, Sanderson JB (2003) Developing disease-suppressive soils through crop rotation and tillage management practices. Soil Till Res 72:181–192. https://doi.org/10.1016/s0167-1987(03)00087-4

Punja ZK, Zhang YY (1993) Plant chitinases and their roles in resistance to fungal diseases. J Nematol 25:526–540

Purnama MI, Prihatna C, Gunawan AW, Suwanto A (2012) Isolasi dan identifikasi secara molekuler *Ganoderma* spp. yang berasosiasi dengan penyakit busuk pangkal batang di kelapa sawit. J Fitopatol Indones 8(1):9–15

Raaijmakers JM, Weller DM (1998) Natural plant protection by 2,4-Diacetylphloroglucinol-producing *Pseudomonas* spp. in take-all decline soils. Mol Plant Microbe Interact 11:144–152. https://doi.org/10.1094/MPMI.1998.11.2.144

Rahman KA, Othman R (2020) Influence of pH levels on disease development in oil palm seedling roots infected with *Ganoderma boninense*. Rhizosphere 13:100181

Ramasamy S (2000) Cross-infectivity and decay ability of Ganoderma species parasitic to rubber, oil palm and tea. In: Ganoderma diseases of perennial crops. CABI Publishing, Oxon

Rao V, Lim C, Chia C, Teo K (2003) Studies on Ganoderma spread and control. Planter 79:367–383

Rees RW, Flood J, Hasan Y, Wills MA, Cooper RM (2012) *Ganoderma boninense* basidiospores in oil palm plantations: evaluation of their possible role in stem rots of *Elaeis guineensis*. Plant Pathol 61:567–578

Rønn R, McCaig AE, Griffiths BS, Prosser JI (2002) Impact of protozoan grazing on bacterial community structure in soil microcosms. Appl Environ Microbiol 68:6094

Ros M, Almagro M, Fernández JA, Egea-Gilabert C, Faz Á, Pascual JA (2020) Approaches for the discrimination of suppressive soils for *Pythium irregulare* disease. Appl Soil Ecol 147:103439

Rousk J, Brookes PC, Bååth E (2009) Contrasting soil pH effects on fungal and bacterial growth suggest functional redundancy in carbon mineralization. Appl Environ Microbiol 75:1589

Rousk J, Bååth E, Brookes PC, Lauber CL, Lozupone C, Caporaso JG, Knight R, Fierer N (2010) Soil bacterial and fungal communities across a pH gradient in an arable soil. ISME J 4:1340–1351

Rusli MH, Idris A, Cooper RM (2016) Evaluation of Malaysian soils for potential suppressiveness of Fusarium wilt of oil palm caused by *Fusarium oxysporum* f.sp. *elaeidis*. J Microb Biochem Technol 8:459–464

Said N, Omar D, Nasehi A, Wong MY (2019) Pyraclostrobin suppressed Ganoderma basal stem rot (BSR), promoted plant growth and induced early expression of beta-1, 3-glucanase in oil palm (*Elaeis guineensis*). J Oil Palm Res 31:248–261

Sambanthamurthi R, Sundram K, Tan YA (2000) Chemistry and biochemistry of palm oil. Lipid Res 39:507–558. https://doi.org/10.1016/S0163-7827(00)00015-1

Schlatter D, Kinkel L, Thomashow L, Weller D, Paulitz T (2017) Disease suppressive soils: new insights from the soil microbiome. Phytopathology 107(11):1284–1297

Schulz G, Schneider D, Brinkmann N, Edy N, Daniel R, Polle A, Scheu S, Krashevska V (2019) Changes in trophic groups of protists with conversion of rainforest into rubber and oil palm plantations. Front Microbiol 10:240

Shariffah-Muzaimah SA, Idris AS, Nur-Rashyeda R, Naidu Y, Zainol Hilmi NH, Norman K (2020) Impact of pre-inoculating soil with *Streptomyces* sp. GanoSA1 on oil palm growth and Ganoderma disease development. Biocatal Agric Biotechnol 29:101814. https://doi.org/10.1016/j.bcab.2020.101814

Sharma AK, Johri BN, Gianinazzi S (1992) Vesicular-arbuscular mycorrhizae in relation to plant disease. World J Microbiol Biotechnol 8(6):559–563

Sharp R (2013) A review of the applications of chitin and its derivatives in agriculture to modify plant-microbial interactions and improve crop yields. Agronomy 3(4):757–793

Siddiqui Y, Surendran A, Paterson RRM, Ali A, Ahmad K (2021) Current strategies and perspectives in detection and control of basal stem rot of oil palm. Saudi J Biol Sci 28(5):2840–2849

Siegel-Hertz K, Edel-Hermann V, Chapelle E, Terrat S, Raaijmakers JM, Steinberg C (2018) Comparative microbiome analysis of a Fusarium wilt suppressive soil and a Fusarium wilt conducive soil from the Chateaurenard region. Front Microbiol 9:568. https://doi.org/10.3389/fmicb.2018.00568

Sigit W, Demetria DH, Yanuar A, Widya KZ, Eko P, Asep U, Susilo ND, Erni M (2016) Tree crop estate statistic of Indonesia 2015–2017: oil palm. Directorate General of Estate Crops, Jakarta

Sime Darby (2014) Palm oil facts and figures. In: Sime Darby plantation: profile and fact sheets. Sime Darby, Kuala Lumpur, pp 1–8

Singh G (1991) Ganoderma-the scourge of oil palms in the coastal areas. Planter 67:421–444

Steyaert R (1980) Study of some *Ganoderma* species. In: Bull. du Jardin Botanique Nat de Belgique/Bull. van de Nationale Plantentuin van België, pp 135–186

Stutz EW, Defago G, Kern H (1986) Naturally occurring fluorescent pseudomonads involved in suppression of black root rot of tobacco. Phytopathology 76:181–185. https://doi.org/10.1094/Phyto-76-181

Subagio A, Foster H (2003) Implications of Ganoderma disease on loss in stand and yield production of oil palm in North Sumatra. Proceedings of the MAPPS Conference, Kuala Lumpur, Malaysia, 18–20 August 2003

Sujarit K, Mori M, Dobashi K, Shiomi K, Pathomaree W, Lumyong S (2020) New antimicrobial phenyl alkenoic acids isolated from an oil palm rhizosphere-associated Actinomycete, *Streptomyces palmae* CMU-AB204T. Microorganisms 8:350

Sullivan P (2001) Sustainable management of soil-borne plant diseases. ATTRA, USDA's Rural Business Cooperative Service https://www.attra.org

Sundram S (2010) Growth effects by arbuscular mycorrhiza fungi on oil palm (*Elaeis guineensis* Jacq.) seedlings. J Oil Palm Res 22:796–802

Sundram S, Meon S, Seman IA, Othman R (2015) Application of arbuscular mycorrhizal fungi with *Pseudomonas aeruginosa* UPMP3 reduces the development of *Ganoderma* basal stem rot disease in oil palm seedlings. Mycorrhiza 25:387–397. https://doi.org/10.1007/s00572-014-0620-5

Susanto A, Sudharto P, Purba R (2005) Enhancing biological control of basal stem rot disease (*Ganoderma boninense*) in oil palm plantations. Mycopathologia 159:153–157

Svenningsen NB, Watts-Williams SJ, Joner EJ, Battini F, Efthymiou A, Cruz-Paredes C, Nybroe O, Jakobsen I (2018) Suppression of the activity of arbuscular mycorrhizal fungi by the soil microbiota. ISME J 12:1296–1307

Tengoua FF, Bakoum C (2008) Pathogenicity of Cameroon strains of *Fusarium oxysporum* f. sp. *elaeidis* – the causal agent of oil palm vascular wilt. Planter 84(985):233–237

Thangavelu R, Loganathan M, Arthee M, Prabakaran M, Uma S (2020) Fusarium wilt: a threat to banana cultivation and its management. CAB Rev 15:1–23

Tian L, Lin X, Tian J, Ji L, Chen Y, Tran LSP, Tian C (2020) Research advances of beneficial microbiota associated with crop plants. Int J Mol Sci 21:1792

Tin HS, Palaniveloo K, Anilik J, Vickneswaran M, Tashiro Y, Vairappan CS, Sakai K (2018) Impact of land-use change on vertical soil bacterial communities in Sabah. Microb Ecol 75:459–467

Tomihama T, Nishi Y, Mori K, Shirao T, Iida T, Uzuhashi S, Ohkuma M, Ikeda S (2016) Rice bran amendment suppresses potato common scab by increasing antagonistic bacterial community levels in the rhizosphere. Phytopathology 106:7. https://doi.org/10.1094/PHYTO-12-15-0322-R

Trillas MI, Casanova E, Cotxarrera L, Ordovàs J, Borrero C, Avilés M (2006) Composts from agricultural waste and the *Trichoderma asperellum* strain T-34 suppress *Rhizoctonia solani* in cucumber seedlings. Biol Control 39:32–38

Turner PD (1981) Oil palm diseases and disorders. Oxford University Press, Oxford

USDA FAS (2012) Malaysia: stagnating palm oil yields impede growth. United States Department of Agriculture Foreign Agricultural Service

USDAFAS (2018) World agricultural production. United States Department of Agriculture Foreign Agricultural Service

Wakefield (1920) Diseases of oil palm in West Africa. Kew Bulletin, pp 306–308

Wei Z, Gu Y, Friman VP, Kowalchuk GA, Xu Y, Shen Q, Jousset A (2019) Initial soil microbiome composition and functioning predetermine future plant health. Sci Adv 5:eaaw0759

Weller DM, Raaijmakers JM, Gardener BBM, Thomashow LS (2002) Microbial populations responsible for specific soil suppressiveness to plant pathogens. Annu Rev Phytopathol 40:309–348

Woittiez LS, van Wijk MT, Slingerland M, van Noordwijk M, Giller KE (2017) Yield gaps in oil palm: a quantitative review of contributing factors. Eur J Agron 83:57–77

Wong WC, Tung HJ, Fadhilah MN, Midot F, Lau SYL, Melling L, Astari S, Hadziabdic Đ, Trigiano RN, Goh KJ, Goh YK (2021) Genetic diversity and gene flow amongst admixed populations of *Ganoderma boninense*, causal agent of basal stem rot in African oil palm (*Elaeis guineensis* Jacq.) in Sarawak (Malaysia), Peninsular Malaysia, and Sumatra (Indonesia). Mycologia 113:902–917

Wu M, Zhang H, Li X, Zhang Y, Su Z, Zhang C (2008) Soil fungistasis and its relations to soil microbial composition and diversity: a case study of a series of soils with different fungistasis. J Environ Sci 20:871–877

Xiao Y, Liu X, Meng D, Tao J, Gu Y, Yin H, Li J (2018) The role of soil bacterial community during winter fallow period in the incidence of tobacco bacterial wilt disease. Appl Microbiol Biotechnol 102:2399–2412. https://doi.org/10.1007/s00253-018-8757-3

Yu F, Chen Y, Huang X, Shi J, Xu J, He Y (2023) Does straw returning affect the root rot disease of crops in soil? A systematic review and meta-analysis. J Environ Manag 336:117673. https://doi.org/10.1016/j.jenvman.2023.117673

Zhang ST, Liu XJ, Zhou KH, Deng LY, Zhao WZ, Liu Y, Ding W (2022) Alleviating soil acidification could increase disease suppression of bacterial wilt by recruiting potentially beneficial rhizobacteria. Microbiol Spectrum 10(2). https://doi.org/10.1128/spectrum.02333-21

Zhao Q, Xiong W, Xing Y, Sun Y, Lin X, Dong Y (2018) Long-term coffee monoculture alters soil chemical properties and microbial communities. Sci Rep 8:6116. https://doi.org/10.1038/s41598-018-24537-2

Zhao Y, Zhang M, Yang W, Di HJ, Ma L, Liu W, Li B (2019) Effects of microbial inoculants on phosphorus and potassium availability, bacterial community composition, and chili pepper growth in a calcareous soil: a greenhouse study. J Soils Sediments 19:3597–3607

Zhou LW, Cao Y, Wu SH, Vlasák J, Li DW, Li MJ, Dai YC (2015) Global diversity of the *Ganoderma lucidum* complex (Ganodermataceae, Polyporales) inferred from morphology and multilocus phylogeny. Phytochemistry 114:7–15

Chapter 7
Rhizobacterial-Plant Interactions: Conferring to Plant Growth Under Drought and Salinity Stress

Vahida Rehman, Samala Manoj Kumar, Vadlamudi Srinivas, Jevenary N. Lukeye, and Subramaniam Gopalakrishnan

1 Introduction

Impact of Environmental Factors on Plant Growth and Productivity Global climate change has a harsh impact on survival of natural ecosystems. It is a dynamic, multifaceted system of changes in environmental conditions that effect both abiotic and biotic components of the planet. Fluctuations in the global climate change have drawn considerable attention from researchers around the world, as these changes have a strong impact on agricultural production and food security worldwide (Watts et al. 2021). A major challenge for agriculture is to feed an increasing human population in the backdrop of production losses provoked by climate change and at the same time mitigate its effects on the environment (Chaudhry and Sidhu 2021). These changes results in alteration in environmental conditions including intensity of rainfall, heat waves, CO_2 concentration, soil salinity, droughts, floods and freezing patterns had reduced crop yields, also leads to emergence of new pests and pathogens (Chaudhry and Sidhu 2021). Thus, changing climate conditions have led to significant rises in global temperatures and the emergence of new abiotic stresses, posing serious challenges to agricultural productivity (Kumawat et al. 2022). Major crops such as wheat, rice, maize and soybean, which contribute to two-thirds of global food supply are predicted to experience yield reductions of 6.0%, 3.2%, 7.4% and 3.1% respectively, for every 1 °C rises in global average temperature

V. Rehman · V. Srinivas
Biocontrol Lab, International Crops Research Institute for the Semi-Arid Tropics (ICRISAT), Patancheru, Hyderabad, India

S. M. Kumar
Department of Microbiology, Tamil Nadu Agricultural University, Coimbatore, India

J. N. Lukeye · S. Gopalakrishnan (✉)
International Institute of Tropical Agriculture (IITA), Dar es Salam, Tanzania
e-mail: s.gopalakrishnan@cgiar.org

S. Gnanamanickam, M. Höfte (eds.), *Disease Suppressive Tropical Soils*,
https://doi.org/10.1007/978-3-032-08100-1_7

(Zhao et al. 2017). High temperatures are associated with various effects, including altered shoot and root growth, changes in biomass production, shifts in flowering time, and reduced seed production, all of which negatively impact crop yields. Drought has also been a significant factor hindering agricultural productivity over the past five decades. Limited rainfall reduces the natural leaching process, where water percolates through the soil, flushing out salts. Without this natural mechanism, salt concentrations build up in the soil, making it increasingly saline. Thus, drought and salinity are the most devastating abiotic stresses, leading to secondary harmful effects. These include oxidative stress and osmotic stress, which are common to both stresses, as well as ionic stress specifically associated with salt stress (Choudhary et al. 2022).

Salinity is a stress of global magnitude, having a substantial effect on plant growth, and is accountable for a significant loss in their productivity. Exerting adverse effects on germination, vigour, and yield, it led to drastic decrease in plant productivity which was seen in plants growing in the arid- and semi-arid regions (Paul and Lade 2014). Salinity stress affects around 20% of cultivable land worldwide, and rising salinity levels cause the loss of 1–2% of arable land each year. (Arora et al. 2020; Noori et al. 2021). A saline soil has pH of 8.5, sodium adsorption ratio (SAR) of 4.0 dS m^{-1} and an exchangeable sodium percentage (ESP) of about 15.0 (Allotey et al. 2008; Etesami 2018). The Food and Agriculture Organization (FAO) estimated that more than 6% of land area has been affected due to salinity (Munns and Tester 2008; Ilangumaran and Smith 2017). More than one-third of the irrigated cropland in the world is impacted by soil salinity. The major contributors to soil salinity are rise in level of groundwater salinity and inadequate drainage systems. The staple crops such as wheat, rice, and corn that are important in human diets are glycophytes and cannot survive in soils with NaCl concentrations greater than 200 mM (Zhao et al. 2020; Kumawat et al. 2022). A water deficit is generated in the roots as osmotic phase is initiated due to salt accumulation in the root system (Munns and Tester 2008; Safdar et al. 2019; Mukhopadhyay et al. 2021). Due to osmotic pressure plant reduces water usage by decreasing the growth rate and stomatal closure (Munns and Tester 2008; Shrivastava and Kumar 2015; Ilangumaran and Smith 2017; Safdar et al. 2019). An increase in the salt concentration disturbs the cellular ionic homeostasis thus leading to an increase in the rate of reactive oxygen species (ROS) production followed by the ion phase which relates to excess of ions primarily Na^+ and Cl^- (Shrivastava and Kumar 2015; Munns and Tester 2008; Dodd and Pérez-Alfocea 2012). The accumulation of Na^+ and Cl^- ions exceeds the rate of exclusion when the salt concentration is high. So, the excess radicles aggregate mainly in the vacuoles and cytoplasm of leaves causing structural and functional changes in the cell (Munns and Tester 2008; Ilangumaran and Smith 2017; Mukhopadhyay et al. 2021). Additionally, the increased salt concentrations can also limit the uptake of macro nutrients (P and N) and micronutrients (Mg, Fe, Cu, and Zn) either by reducing their solubility or by Na^+ and Cl^- competing with the uptake of nutrients (Bano and Fatima 2009; Talei et al. 2012). Consequently, ion toxicity and osmotic stress can cause photosynthetic imbalances leading into oxidative stress (Munns and Tester 2008; Shrivastava and Kumar 2015; Ilangumaran and Smith 2017). These metabolic changes increase the rate of the formation of ROS, by-products that affect cellular components, such as photosystems, and which can induce programmed cell death (Apel and Hirt 2004; Munns and Tester 2008; Ilangumaran and Smith 2017; Safdar et al. 2019). Thus, various factors contributing to plant growth and development such as germination, nodulation, vegetative growth, water and nutrient uptake, reproductive development and nitrogen fixation are disrupted due to these effects (Shrivastava and Kumar 2015; Chandra et al. 2021; Prajapati et al. 2022). Salinity induces osmotic stress which can be considered as a form of a physiological drought. However, at later stages of plant growth leaf senescence is induced in plants due to high accumulation of salts in transpiring leaves leading to ionic toxicity (Munns and Tester 2008). Thus, the complex relationship between components of drought and salinity stress result in secondary stresses such as osmotic and oxidative stress (during drought and salinity) as well as ionic stress (during salinity) leading to demise of plant (Gill and Tuteja 2010).

Drought is one of the stresses limiting agricultural productivity and has a significant impact on crop yield (Khan et al. 2013). Due to changes in precipitation patterns and evapotranspiration caused by global warming there has been an increase in the frequency and severity of drought stress (Dai 2011). Drought occurs when there is not enough water in the soil for plant roots to absorb. It is recognized as a multifaceted stress that affects plants at the morphological, phenological, and molecular level (Kaur and Asthir 2017). Relative water content (RWC), stomatal conductance, leaf water potential, leaf temperature, rate of transpiration, and canopy temperature are affected under significant water scarcity (Ings et al. 2013). Drought stress also reduces the photosynthesis activity by reducing the leaf area and hinders mitosis, elongation and expansion of cells which limits growth (Potopova et al. 2016). Drought hinders grain filling and flower formation by slowing plant growth and development. Eventually, this results in reduced yield (Kaur and Asthir 2017). Drought also significantly affects structures in tissue, cells and subatomic levels. Significant alterations occur in chloroplast and vacuole structures at the subcellular level. For example, stromal lamellae enlarge, changing the structural organization of thylakoids and leading to the creation of antenna chlorophyll depleted photosystem (PS) II, which reduces photosynthetic and respiratory activity (Kaur and Asthir 2017). The impacts attributed by increased salinity and drought in agriculture at global scale necessitate the need to have alternatives to curb these agricultural production constraints.

Potential of PGPR to Sustainably Enhance Stress Tolerance Therefore, finding sustainable solutions for crop adaptation to changing climatic conditions and to enhance crop production is key to guarantee food security around the globe (Sánchez-Bermúdez et al. 2022). The integration of beneficial microorganisms and/or bioproducts into agricultural practices should be promoted to enhance crop productivity sustainably and environmentally friendly under environmental stress situations (Gill et al. 2016). Plant growth-promoting rhizobacteria (PGPR) are vital microorganisms with a unique ability to promote plant health in a variety of ways, to maintain the plants' life in the rhizosphere and to encourage the biological processes that impact the plants' development and survival (Kumar et al. 2019a). PGPR enhances soil-water-plant relations through intricate mechanisms and subtle signalling cues that are not yet well-understood. Currently, new data from molecular studies has provided information regarding the signalling networks of plant–microbe interactions which assist in salt and drought tolerance (Ilangumaran and Smith 2017; Kumar et al. 2019b; Danish et al. 2020). PGPR plays prominent role in enhancing important physiological processes like photosynthesis, nutrient uptake, source-sink relationships and water uptake that stimulate growth and development of plant (Gopalakrishnan et al. 2015, 2022). It also employs a combination of morphological, physiological and molecular strategies to help plants adapt to salinity stress and maintain growth and productivity. Exploiting these beneficial plant-microbe interactions is a promising approach for sustainability in agriculture.

2 How PGPR Promote Salt and Drought Tolerance?

Plant growth promoting rhizobacteria (PGPR) commonly known as rhizobacteria are the complex assemblage of bacteria inhabiting the region around plant roots, contribute significantly to controlling the negative impacts of salinity and drought on plant growth and development. These microorganisms are capable of beneficial interactions with the plants, which makes possible for the plants to adapt to and flourish in saline conditions. PGPR assist in restoring balance of ions and osmotic levels to prevent cell injury and enable growth in high salinity environments (Otlewska et al. 2020; Liu et al. 2020). PGPR employ various mechanisms to alleviate negative impacts of salinity stress on plants. These mechanisms encompass osmotic adjustment by producing compatible solutes such as betaines, trehalose and proline which helps to maintain water balance and cellular homeostasis. Certain PGPR can regulate the uptake and distribution of essential nutrients, while also reducing the accumulation of toxic ions like sodium, thereby maintain the optimal ionic balance within the plant (Fig. 7.2; Table 7.1). They can also stimulate the production of various antioxidant enzymes and compounds in plants, effectively neutralizing the harmful effects of reactive oxygen species generated during salinity stress (Kumawat et al. 2022; Prajapati et al. 2022). Some PGPR are capable of synthesizing or regulating the levels of plant growth hormones, such as auxin, cytokinin and gibberellin which can enhance the plants resilience and adaptive responses to salinity in plants which brings out sustainable crop yield and soil health (Gopalakrishnan et al. 2012a, b, 2013a, b, 2014, 2015a, b, c, 2016a, b, c; Srinivas et al. 2020, 2022). Some PGPR can form protective biofilms around plant roots, creating a micro-environment that shields plant from detrimental effects of salinity stress. Also, PGPR employs several mechanisms such as elongation of root and shoot, changes to osmotic tuning, release of plant hormone levels, stomatal conductivity, increasing nutrient and water uptake, volatile organic compound (VOC) release, and exo-polysaccharide production (EPS) by which microbes exhibit resistance to drought stress to protect plant (Fig. 7.1; Table 7.2).

Table 7.1 Mechanisms of PGPR-plant interactions alleviating salinity stress tolerance

Organism	Type of stress	Mechanism	Crop	References
Pseudomonas sp. *P8, Peribacillus* sp. *P10, and Streptomyces* sp. *X52*	Salinity	Salt resistance ability, secrete osmo-protectants, accelerated plant growth	*Zea mays*	Peng et al. (2021)
Enterobacter cloacae	Salinity	Upregulation of stress related genes SOD and APX; EPS production, ACC deaminase production	*Zea mays*	Ali et al. (2022)
Brevibacterium sediminis	Salinity	Could tolerate 12% NaCl stress	*Oryza sativa*	Mahmud-Ur-Rahman Naser et al. (2022)
Azospirillum sp.	Salinity	Improved phosphate solubilization, plant growth due to reduced salinity stress	*Cicer arietinum*	Patel et al. (2012)
Pseudomonas putida	Salinity	Reduced salinity stress, improved plant growth	*Cicer arietinum*	Nagpal et al. (2019)
Pseudomonas fluorescens, P. putida, and Bacillus subtilis	Salinity	Enhanced osmolytes and POD synthesis	*Glycine max*	Abulfaraj and Jalal (2021)
Acinetobacter bereziniae, Enterobacter ludwigii, and Alcaligenes faecalis	Salinity	Enhanced osmolytes soluble sugar and proline synthesis	*Pisum sativum*	Sapre et al. (2021)
Glomus mosseae	Salinity	Increased CAT, SOD, APX, and POD activity in roots. Reduced cell membrane damage and increased plant growth	*Solanum lycopersicum*	He et al. (2007)
Bacillus subtilis	Salinity	Transcriptional changes in ion transporter genes of HKT type	*Puccinellia tenuiflora*	Niu et al. (2016)
Bacillus atrophaceus ZYH01 *Planococcus soli WZYH02*	Salinity	Increased K^+/Na^+ ratio, regulates NHX and HKT genes	*Zea mays*	Hou et al. (2022)
Pseudomonas oryzihabitans AXSa06	Salinity	Increased ascorbate and MDA content, peroxidase and catalase activity	*Solanum lycopersicum*	Mellidou et al. (2021)
Bacillus sp.	Salinity	Induces plant response for defence enzymes, chlorophyll, proline and soluble sugar under salt stress	*Zea mays*	Misra and Chauhan (2020)
Thalassobacillus denorans (NCCP-58), *Oceanobacillus kapialis* (NCCP-76)	Salinity	Enhanced germination and growth of root and shoot, developed pigment system, reduced Na + ion accumulation	*Oryza sativa*	Shah et al. (2017)

(continued)

Table 7.1 (continued)

Organism	Type of stress	Mechanism	Crop	References
Dietzia natronolimnaea	Salinity	Increased expression of SOS related genes, tissue-specific expression of ion transporters, modulation of ABA signalling cascade	*Triticum aestivum*	Bharti et al. (2016)
Arthrobacter spp. SU18, *B. aquimaris* SU44, *B. aquimaris* SU8	Salinity	Enhanced root dry weight and shoot biomass under salinity stress conditions	*Triticum aestivum*	Upadhyay and Singh (2015)
Bacillus aryabhattai	Salinity	Exo-polysaccharide production	*Oryza sativa*	Sultana et al. (2020)
Enterobacter sp.	Salinity	Enhanced production of SOD, CAT and GSH	*Oryza sativa*	Sagar et al. (2020)
Pseudomonas azotoformans FAPS	Salinity	Biofilm formation which improved physiological and morphological attributes	*Zea mays*	Ansari et al. (2021)
Paenibacillus mucilaginosus	Salinity	Reduced the Na^+ ions toxicity in shoot and root through VOCs production. Increased proline	*Glycine max*	Ma et al. (2018)
Bacillus megaterium	Salinity	Increased production of phytohormones and polyamines, which increased root, shoot and leaf growth	*Solanum lycopersicum*	Nascimento et al. (2020)
Bacillus methylotropicus, B. siamensis, Bacillus sp.	Salinity	ACC deaminase production	*Triticum aiestivum*	Amna et al. (2019)
Aneurinibacillus aneurinilyticus and Paenibacillus sp.	Salinity	ACC deaminase activity	*Phaseolus vulgaris*	Gupta and Pandey (2019)
Bacillus amyloliquefaciens	Salinity	Biosynthesis of VOCs	*Arabidopsis thaliana*	Liu et al. (2020)
Halomonas maura, Ensifer meliloti	Salinity	EPS production, increased plant biomass and leghaemoglobin activity, Nodulation.	*Medicago sativa*	Martínez et al. (2016)
Burkholdera cepacian, Promicromonospora sp. *& Acinetobacter calcoaceticus*	Salinity	Downregulation of ABA genes, increases plant biomass, K^+ uptake, chlorophyll content	*Cucumis sativus*	Kang et al. (2014a)
Kocuria rhizophila	Salinity	Transcriptional changes in ion transporter genes (HKT & NHX type) and hormonal changes (IAA & ABA), reduces Na^+ accumulation.	*Zea mays*	Li et al. (2020)
Pseudomonas putida	Salinity	Phytohormonal changes of IAA and ABA levels, enhanced germination rate, nutrient content	*Gossypium hirsutum*	Yao et al. (2010)

(continued)

Table 7.1 (continued)

Organism	Type of stress	Mechanism	Crop	References
Bacillus sp. BH32	Salinity	Upregulation of *ipdC* gene for indole 3 pyruvate decarboxylase production for plant growth production	–	Belaouni et al. (2022)
Pseudomonas thivervalensis SC5	Salinity	Upregulation of acdS gene for 1-aminocyclopropane-1-carboxylate deaminase production.	–	Nascimento et al. (2021)
Bacillus subtilis GB03	Salinity	Upregulation of PtHKT1,5 and PtSOS1 genes, and downregulation of PtHKT2,1 which modulates Na^+ homeostasis by K^+ absorption	*Puccinellia lenuiflora*	Kang et al. (2019)

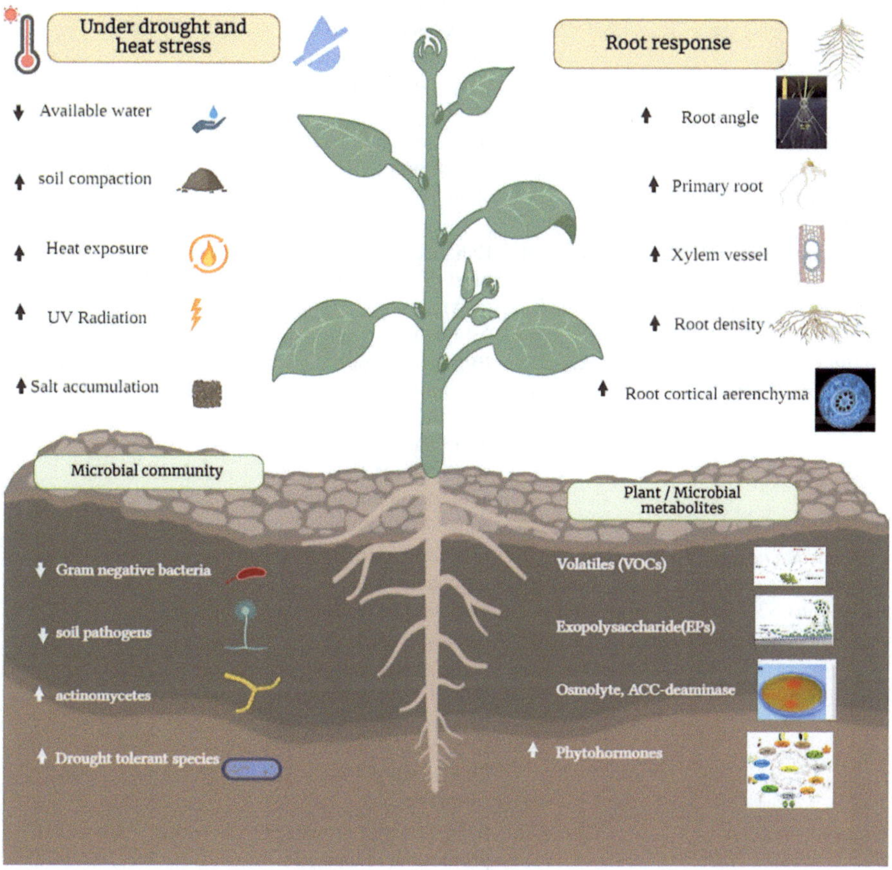

Fig. 7.1 PGPR mechanisms tolerating drought stress

Table 7.2 Mechanisms of PGPR-plant interactions alleviating drought stress tolerance

Organism	Stress	Mechanism	Crop	Reference
Bacillus pumilus DH11, *Bacillus firmus* 40	Salt/ Drought/ HMs	ACC deaminase, siderophore production, Phosphorous solubilization, proline production	*Solanum tuberosum*	Gururani et al. (2013)
Achromobacter xylosoxidans (SF2), *Bacillus pumilus* (SF3 and SF4)	Drought	High salicylic acid content, Abscisic acid production.	*Helianthus annuus*	Castillo et al. (2013)
Azospirillum brasilense SP-7	Drought	Increased osmoregulation, reduced ABA levels; downregulation of ZmVP14 gene	*Zea mays*	Cura et al. (2017)
Variovorax paradoxus 5 C-2	Drought	Reduces ethylene production, enhanced nodulation, xylem ABA concentration, seed nitrogen content, water content and yield	*Pisum sativum*	Belimov et al. (2009)
E. cloacae, P. aeruginosa, lactobacillus adecarboxylata + biochar, *Azotobacter xylosoxidans*	Drought	Increased grain yield and chlorophyll content	*Zea mays*	Danish et al. (2020)
Pseudomonas sp. RJ15, *B. subtilis* RJ46, *Ochrobactrum pseudogrignonense*	Drought	Enhanced the activity of ROS-quenching enzymes and osmolytes	*P. sativum* and *V. mungo*	Saikia et al. (2018)
Bacillus cereus (ESD3, ESD21 and ESB22)	Water deficient stress	EPS production, biofilm formation, IAA, ACC deaminase, siderophores, antioxidant enzymes, enhanced biomass	*Solanum lycopersicum*	Haque et al. (2020)
Cupriavidus necator 1C2 (B1) *and Pseudomonas fluorescens* S3X (B2),	Water deficient stress	Increased root/shoot ratio, enhanced biomass production	*Zea mays*	S.I.A. Pereira (2020)
Pseudomonas azotoformans	Drought	Increased seed germination rate, shoot length and root length, produced EPS, ACC deaminase, IAA, biofilm formation.	*Triticum aestivum*	Ansari et al. (2021)
Glucanacetobacter diazotrophicus Pal5	Drought	Increased root growth promotion, enhanced osmo-protectant solutes	*Oryza sativa*	Silva et al. (2020)

(continued)

Table 7.2 (continued)

Organism	Stress	Mechanism	Crop	Reference
Bacillus megaterium MU2	Drought	Increased chlorophyll content (a, b), carotenoids, relative water content (RWC)	*Triticum aestivum*	Rashid et al. (2022)
Bacillus subtilis HAS31	Drought	Improved chlorophyll content, RWC, osmolytes production, antioxidants enzymes, shoot growth, yield	*Solanum tuberosum L.*	Batool et al. (2020)
Leclercia decarboxylata, Agrobacterium fabrum	Drought	High chlorophyll contents, enhanced nutrient uptake	*Triticum aestivum*	Danish et al. (2019)
Pseudomonas fluorescens DPB15 *and P. palleroniana* DPB16	Drought	Enhanced shoot and root growth	*Triticum aestivum*	Chandra et al. (2018)
Bacillus thuringiensis	Drought	Increased photosynthesis rate, biomass production, enhanced volatiles emission,	*Triticum aestivum*	Timmusk et al. (2014)
P. aeruginosa PM389, *P. aeruginosa* ZNP1, *B. endophyticus* J13 *and B. tequilensis* J12	Drought	Produced IAA, cytokinin, gibberellins, EPS secretion	*Arabidopsis thaliana*	Ghosh et al. (2019)
Pseudomonas fluorescens, Enterobacter hormaechei, Pseudomonas migulae	Drought	Increased seed germination and seedling growth	*Setaria italica*	Niu et al. (2018)
O. pseudogrignonens eRJ12	Drought	Reduced ACC deaminase activity	*Pisum sativum*	Saikia et al. (2018)
Bacillus subtilis, paenibacillus illinoinensis	Drought	Enhanced root length, transpiration, photosynthetic rate, proline content	*Capsicum annum*	Vigani et al. (2019)
Bacillus subtilis GOT9	Drought	Development of lateral roots, drought stress tolerance, growth	*Arabidopsis, Canola*	Woo et al. (2020)
Trichoderma sp. *Pseudomonas* sp.	Drought	Over expression of water permeability (OSPiP), drought adaptation (DHN) and dehydration genes (DREB)	*Oryza sativa*	Singh et al. (2020)
Bacillus subtilis FAB1, *Pseudomonas azotoformans*-FAP3	Drought	Biofilm formation, increased root and shoot length	*Triticum aestivum*	Ansari et al. (2023)

3 PGPR Mechanisms Alleviating Salinity Stress

Ion Homeostasis: Selective Ion Uptake Salinity inhibits growth of plant because of increased Na^+ concentration and low K^+ /Na^+ ratio in plant. During saline stress plants can protect themselves by restricting Na + entry, expelling Na^+ from roots and retrieving Na^+ from xylem stream for recirculation to roots (Kaushal and Wani 2016). To combat the higher concentration of Na^+ in the environment, plants must reduce influx of Na^+ through the plasma membrane. The influx of Na^+ is mediated by the high-affinity K^+ channel (HKT), which plays critical roles in influx of Na^+ across the plasma membrane and long-distance Na^+ transport, which is a key determinant of plant salt tolerance (Liu et al. 2022). With localization on the plasma membrane, the Na^+/H^+ antiporter (also referred to as SOS1 salt overlay sensitive channel) efflux Na^+ in response to its increasing cytosolic levels (Qiu et al. 2002). Also, the increase in plant Na^+ level interferes with K^+ ion uptake at the root surface *via* the low-affinity K^+ uptake system. Numerous PGPR have been reported to possess the ability to enhance the shoot and/or root K^+ /Na^+ ratios by raising K^+ and/or lowering Na^+ levels in the cells, but the exact mechanisms differ, which in turn limits their accumulation in the upper parts of the plants, thus aiding in the prevention of disturbance in ionic equilibrium in plants. Also, promoting biofilm formation at the root surface that prevents the influx of Na^+ into the roots, EPS production by PGPR strains traps cations in their matrix, making them unavailable for uptake by the plants (Dodd and Pérez-Alfocea 2012). VOC produced by *Bacillus subtilis* reduced the transcriptional expression of a high-affinity K^+ transporter (HKT1) in the roots of *Arabidopsis*, while increasing its expression in the shoots. This led to decreased Na^+ uptake by the roots and enhanced Na^+ expulsion from the shoots. Maize plants inoculated with an ACC deaminase (ACCd)-producing rhizobacterial strain showed increased uptake of nitrogen (N), phosphorus (P), and potassium (K), resulting in a higher K^+/Na^+ ratio, which enhanced growth in saline environments (Nadeem et al. 2009). Similarly, inoculating pea plants with the ACCd-containing rhizobacterium, *Variovorax paradoxus* 5C-2 enhanced nodulation, leading to increased seed nitrogen content in drying soil. Ethylene inhibits nodulation, but by reducing drought-induced increases in ACC content in xylem sap, rhizobacteria promoted plant growth during drought conditions (Belimov et al. 2009). Another PGP bacteria, *Pseudomonas simiae* reported by Vaishnav et al. (2015) had shown transcriptional changes in accumulation of proline, phosphatase activity and volatile organic compounds production which increased weight, length and K^+ /Na^+ ratio in soybean crop. Similarly, *Azospirillum brasilense and Azotobacter chroocсum* had shown transcriptional changes in accumulation of polyphenols and plant growth genes which leads to higher plant growth and ion uptake in *Zea mays* L. (Hamdia et al. 2004; Rojas-Tapias et al. 2012).

Synthesis of Stress Relief Compounds: Phytohormones and ACC Deaminase Production Production of phytohormones by various PGPR strains can modulate plant hormones, metabolites and enzymes, thus enhancing plant growth under stressed conditions (Kaushal and Wani 2016). Phytohormones such as abscisic acid (ABA) or indole-3-acetic acid (IAA), or ethylene are synthesized to alleviate salt stress by plants. These hormones cause modification in plant morphology, metabolism, production of biomass, and other mechanisms. Thus, this mechanism of hormonal signalling changes due to plant and PGPR interactions can induce soil salinity tolerance in plants (Vaishnav et al. 2016; Ayuso-Calles et al. 2021). Among various compounds, most common and widely studied bacterial signalling molecules in plant–microbe interactions are IAA. The endogenous IAA levels in plants have a prominent role in the function of exogenous IAA (Ilangumaran and Smith 2017). Several studies have reported that foliar application of exogenous IAA to various crops (sorghum, maize, and pea) mitigated salt stress by modulation of Na^+ accumulation, weight measurements, water use efficiency, or photosynthesis rate (Azooz et al. 2004; Groppa et al. 2012; Bhattacharyya et al. 2015). The application of *Bacillus amyloliquefaciens* SQR9 significantly enhanced the salt stress tolerance (100 mM NaCl) in maize seedlings grown under *in vitro* conditions. Bacterial inoculation led to increased chlorophyll content and levels of total soluble sugar, boosted peroxidase and catalase activities, elevated glutathione content, and improved the K^+/Na^+ ratio. Additionally, the salinity-induced rise in abscisic acid levels was mitigated by SQR9 inoculation, maintaining abscisic acid (ABA) at normal levels. These physiological adaptations to alleviate salt stress were further supported by the upregulation of genes such as RBCS and RBCL (encoding RuBisCo subunits), H (+)- Ppase (encoding H^+ pumping pyrophosphatase), HKT1, NHX1, NHX2, and NHX3, alongside the downregulation of NCED expression (encoding 9-cis-epoxycarotenoid dioxygenase) in the inoculated seedlings (Chen et al. 2016). *Enterobacter* sp. EJ01, isolated from the halophyte sea China pink (*Dianthus japonicus* Thunb), enhanced plant growth and salt stress tolerance (200 mM NaCl) in both *Arabidopsis* and tomato plants. A short-term treatment (6 hours) with EJ01 led to increased expression of genes linked with salt stress response in *Arabidopsis* seedlings, including DRE-binding proteins (DREB2b), desiccation-related genes (RD29A, RD29B), late embryogenesis abundant (LEA) genes (RAB18), proline biosynthesis genes (P5CS1 and P5CS2), and stress-inducible priming genes (MPK3 and MPK6). Additionally, GFP-tagged EJ01 was observed colonizing both the rhizosphere and endosphere (interior of roots) of *Arabidopsis* roots. In tomato plants, inoculation with EJ01 under salt stress conditions resulted in enhanced reactive oxygen species (ROS) scavenging activities, including increased activity of the antioxidant enzyme ascorbate peroxidase (Kim et al. 2014).

ABA is recognized as a stress hormone and is upregulated in salinity stress (et al. 2020). ABA assists in the accumulation of osmolytes, Ca_2^+ and K^+, modulates cell ion balance, prompts the roots elongation and the appearance of lateral roots, increases old leaf shedding, and controls leaf stomatal closure (Bhat et al. 2020; et al. 2020; Nian et al. 2021). The ABA pathway plays a crucial role in helping plants manage salt stress. When plants experience osmotic stress due to high salt levels, ABA accumulates, triggering an acclimation response. ABA influences various plant functions, such as seed germination, cell differentiation, stomatal movement, production of stress-related proteins and metabolites and organ development. These vital processes for minimizing the damage initiated by stress in plants (Song et al. 2022). The halotolerant bacterium *Dietzia natronolimnaea* STR1 enhanced salinity tolerance (150 mM NaCl) in wheat plants by modulating an ABA-signalling cascade. This was evidenced by the upregulation of the ABA-responsive gene *TaABARE* and the 12-oxophytodienoate reductase 1 gene *TaOPR1*, which in turn activated *TaMYB* and *TaWRKY*. This activation led to expression of stress response genes, including upregulation of the salt stress-induced gene *TaST*. *Bacillus pumilus* MAK9 enhanced salinity stress tolerance in Chinese cabbage and the endophytic bacterium *Curtobacterium* sp. SAK1 enhanced salinity tolerance in soybean, along with plant overall development (Khan et al. 2019a). When soybean plants grown in 300 mM NaCl were inoculated with *Arthrobacter woluwensis* AK1 upregulation in the expression of stress-resistance genes (*GmST1* and *GmLAX3*) was observed due to synthesis of phytohormones such as IAA and ABA (Khan et al. 2019b). The treatment with *Bacillus subtilis* IB22 increased the levels of ABA in roots while reducing levels of ABA in shoots of barley plants grown under a salt concentration of 100 mM. This effect was achieved by upregulating the expression of gene *HvNCED2*, which is involved in ABA biosynthesis, and downregulating the gene *HvCYP707A1*, which is responsible for ABA degradation.

Ethylene is produced through ACC oxidase (ACCO) from ACC (Oleńska et al. 2020; Shilev 2020). When ACCO synthesizes ethylene higher than its threshold amount can cause "stress ethylene" in the plant. Plants produce "ethylene" under stress conditions such as salinity, drought, or any other abiotic stress which primarily limits plant growth. Ethylene is a vital hormone for plant growth and development at low concentrations. However, under stress, it is overproduced, affecting the growth and yield of all plants exposed to stress (Prajapati et al. 2022). Therefore ACC (1-aminocyclopropane-1-carboxylic acid) deaminase, an important bacterial physiological characteristic, aids plant growth under stress situations (Etesami et al. 2015). PGPR with the enzyme ACC deaminase would lower the level of stress ethylene and allow the plant to flourish under abiotic stress conditions (Forni et al. 2017). In stressed plants, PGPRs containing ACC deaminase can convert ACC, a plant ethylene precursor, to ammonia and -ketobutyrate, lowering ethylene synthesis (Glick 2014). ACCD (1-aminocyclopropane-1-carboxylate deaminase) produced by PGPR can counteract the inhibitory effect of elevated ethylene levels on auxin (IAA) production in plants. These ACCD-producing microbes can regulate the levels of ACC (1-aminocyclopropane-1-carboxylate), the immediate precursor of ethylene in the rhizosphere, phyllosphere and within plant tissues such as the endosphere and root nodules. By utilizing the ACC released by plants or present in these environments, the ACCD-producing microbes effectively limit the availability of ACC, thereby reducing ethylene production. This reduction in ethylene levels helps alleviate stress and promote plant growth under challenging conditions such as those caused by environmental stresses or nutrient deficiencies. *Enterobacter cloacae*, isolated from *Ziziphus nummularia*, improved wheat growth by increasing ACCD activity under salinity stress (200 mM). Similarly, the endophytic *Paenibacillus xylanexedens* PDR6, found in *Phoenix dactylifera* roots, exhibited ACCD activity and reduced ethylene levels under salt stress (Singh et al. 2022a, b). Reduced ethylene production, facilitated by ACCD activity, alleviated salt stress in red pepper and tomato plants when inoculated with *Pseudomonas* sp. and *P. azotoformans* CHB 1107, respectively (Gamalero and Glick 2022). The ACC deaminase (ACCD)-producing halo-tolerant *Brevibacterium linens* RS16 enhanced growth traits in rice and red pepper plants induced with salt stress (Chatterjee et al. 2018).

Biofilm Formation Complex interactions between plants and microorganisms are increasingly recognized as critical determinants of plant health and growth (Berendsen et al. 2012; Mauchline and Malone 2017). In extreme environments, relationships among organisms can often be exclusionary and governed by intricate mechanisms. The system that includes soil subjected to salt stress, plants, and PGPR creates a distinctive extreme environment that fosters mutualistic relationships. In this context, interactions between plants and microbes are vital for triggering or enhancing various adaptive mechanisms that enable survival in conditions of salt stress (Nadeem et al. 2014; Acosta-Motos et al. 2017; Otlewska et al. 2020). Therefore, halophilic PGPR produces important biological molecules, such as phytohormones and exopolysaccharides (EPS), being crucial in plant–microbiome interaction, in keeping the stability of the soil structures and water-holding capacity and are also useful for bioremediation in saline environments. The biofilm formation and EPS production plays a crucial role in facilitating interactions between plants and microbes by creating a microenvironment that allows these microorganisms to thrive under stressful conditions (Petrić et al. 2022). Under salinity conditions, EPS production alone accounts for 40–90% of the weight of the bacterial extracellular matrix (Li et al. 2018; et al. 2023). EPS facilitate microbial adhesion to plant roots, aid in biofilm formation to prevent cells from drying out due to salt stress and increases the movement of bacteria that are connected to plant roots (Liu et al. 2017; Ruppel et al. 2013; et al. 2023). Bacteria produce polymeric biofilms on a variety of surfaces such as roots and soil, cementing particles, and forming aggregates. This can improve crop performance and soil physicochemical properties. Salt-tolerant strains of *Halomonas* sp. and *Planococcus rifietoensis* strains were reported to improve aggregation of soil and plant growth. These strains form biofilm and accumulate exopolysaccharides in response to increased salt stress (Qurashi and Sabr 2012). Bacterial EPS can mitigate salinity stress by reducing the availability of Na^+ content for plant uptake (Upadhyay et al. 2011). The kind of EPS produced by *Pseudomonas* spp. (Kasotia et al. 2016) and *Bacillus* spp. (Amna et al. 2019) helped in the binding of free Na^+ from the soil, thus making Na^+ unavailable to the soybean and wheat plants, respectively. Reports suggest that guanine cyclises formation is triggered in cells further leading to production of EPS under stress conditions (Borlee et al. 2010). In *A. brasilense* Sp7, *noeJ* and *noeL* genes encoding mannose-6-phosphate isomerise and GDP-mannose 4, 6-dehydratase respectively, are involved in synthesis of EPS (Lerner et al. 2009). The enzymes encoded by these

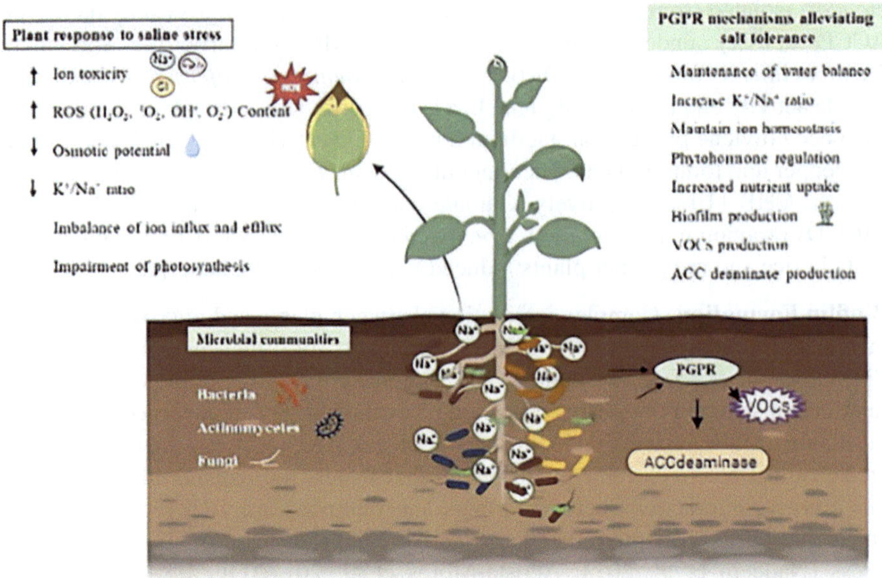

Fig. 7.2 PGPR mechanisms tolerating salinity stress

genes are involved in synthesis of fructose and mannose which are constituents of EPS in *A. brasilense* (Vanbleu et al. 2005). Significant research has been conducted on elevated levels of EPS produced by PGPR in alleviating stress conditions in plants. However, further detailed research must be done to explore the variations in composition of EPS under various stress conditions as each polysaccharide constituent has different structure and water retaining capacity (Kaushal and Wani 2016). Figure 7.2 display various ways PGPR uses to modulate or protect plants from salinity stress.

Stress Gene Regulation PGPR induce and up-regulate the gene expression linked with abiotic stresses tolerance in plants (Khan et al. 2020b; Sharma et al. 2021). Stress-responsive gene expression is mainly regulated *via* transcription factors and prompted through stress sensors that are exposed to complex control of phytohormones (Khan et al. 2020a). PGPR can augment tolerance of the plant to salt stress by regulating the expression of SOS genes, RAB18 (LEA), WRKY TFs, DRE (dehydration responsive element), RD29A, RD29B regulons of ABRE (ABA-responsive elements), besides transcription factor DRE binding proteins (DREB2b) (Chakdar et al. 2019; Etesami 2020). Many stresses induced reactions in plants are controlled by WRKY TFs. However, the role of WRKY Tfs in response to biotic stresses is widely explored whereas studies in abiotic stresses are limited. Studies reported that under saline conditions upregulated expression WRKY28 genes was observed in PGPR-inoculated plants which might have provided tolerance under salt stress conditions (Yu et al. 2016; Wu et al. 2017; Safdarian et al. 2019).

WRKY Tfs also have a prominent role in water and drought stress regulation by scavenging ROS mechanism, expressing different stress-related genes and modulating the cellular osmotic balance as enhance plant growth was reported in PGPR-inoculated plants under greenhouse (Safdarian et al. 2019). *Enterobacter* spp. upregulates salt stress-responsive genes including RAB18, DREB2b, RD29A, and RD29B in *Arabidopsis* in the salinity stress. TaMYB and TaWRKY genes expression are induced in inoculated wheat by *Dietzia natronolimnaea* (et al. 2020). *Bacillus amyloliquefaciens* SN13 upregulates SOS1, SERK1, EREBP, and NADP-Me2 genes and enhances the salt tolerance of rice (Chakdar et al. 2019).

4 PGPR Mitigating Drought Stress

PGPR-Induced Root and Shoot Growth Plant and PGPR interactions are necessary for improved nutrient and water absorption. It has been demonstrated that PGPR increase root length and root density, which improves agricultural output, enables the nutrient uptake and water uptake, and preserves the host plant's osmotic balance by adjusting the relative water content of the plant (Vurukonda et al. 2016). Root exudates contain various high and low molecular weight substances i.e. a significant quantity of lipids, carbohydrates, polysaccharides, and amino acids, which feed microorganisms and boost the microbial populations in the vicinity of the roots (Backer et al. 2018). Changes in the structure of root zone brought by the mutual interaction between plants and PGPR enhances the drought tolerance. The root system architecture is designed to adapt to drought conditions. To absorb more water, plants modify the biomass ratio between their roots and shoots to adjust their root development to drought stress. Therefore, a deeper and broader root system boost plant yield during drought. Under stress plants employ the root to extend the area that is in contact with the moisture in the soil, increasing the hydraulic conductivity. Consequently, the root system improves the water uptake and nutrient uptake, enabling plants to endure drought stress (Farooq et al. 2009).

Bacterial Extracellular Polysaccharides or Exopolysaccaride (EPS) Production The extracellular polysaccharides produced by beneficial bacterial includes cyclic β-glucans, lipopolysaccharides (LPSs), capsular polysaccharides (CPSs), and exopolysaccharides (EPSs) (Perret et al. 2000). These polysaccharides are frequently built up on cell surfaces including a variety of macromolecules that are advantageous to plant growth. EPS play a range of roles in the interactions between microorganisms and plants during drought stress. They are present in a wide range of complex structures with higher quantities of proline, free amino acids, sugars, ~~and~~ monosaccharides and polysaccharides (Morel et al. 2013). Some polysaccharides provide hydrated microenvironment because certain polysaccharides may have higher capacity to retain water than their bulk. In particular, the generation of exopolysaccharides is necessary for plant invasion, nodule development, biofilm formation, surface attachment, root colonization, cell aggregation, and soil aggregation. These processes provide protection against stress to the plant (Sandhya et al. 2009, Fischer et al. 2003; Bomfeti et al. 2011).

Hormonal Modulation Phytohormones plays a vital role in plant adaptation to drought stress. Hormones such as ABA, IAA, brassinosteroids (BR), cytokinin (CK), ethylene, jasmonic acid (JA), and salicylic acid (SA) are involved in the plant reaction to drought (Camaille et al. 2021). PGPR exude phytohormones such as auxins, cytokinins and jasmonates which were reported to enhance various plant growth promoting aspects like root length, number of root tips, shoot elongation, plant fresh weight and plant dry weight, increased chlorophyll content and photosynthetic rate (Yadav and Yadav 2018; Nammala and Smith 2021). Auxins and IAA stimulate division, expansion, and differentiation of cells in plant roots in addition to lateral root development, which aid in the water retention and nutrient uptake of plants (Ilangumaran and Smith 2017). *P. chlororaphis* TSAU13 treated wheat seedlings, cucumber and tomato plants exhibited increased root growth under saline conditions due to enhanced IAA production which increases water conductance in comparison to uninoculated plants (Egamberdieva and Kucharova 2009; Egamberdieva 2012). Additionally, when wheat seedlings were inoculated with *Azospirillum* enhanced IAA synthesis was observed due to up-regulation of indole-3-pyruvate decarboxylase gene resulting in morphological changes in the coleoptile xylem architecture. This adaptation led to increased water conductance in coleoptiles, enabling plant to cope with osmotic stress (Pereyra et al. 2012). When the wheat seedlings grown under drought stress were treated with *Bacillus thuringiensis* which had the ability to produce IAA and 1- aminocyclopropane-1-carboxylate deaminase (ACCd) resulted in increased lateral root density and length by 59%, 200% (Timmusk et al. 2014). The process of germination, elongation of stem, flowering, sex expression and senescence are affected by endogenous gibberellin (Gas) (Kaushal and Wani 2016). Cucumber plants treated with PGPR (*Acinetobacter calcoaceticus* SE370, *Burkholdera cepacia* SE4, and *Promicromonospora* spp. SE188) exhibited elevated endogenous GAs compared to control plants resulting in enhanced plant growth under both drought stress and salinity stress (Kang et al. 2014a). When soybean plants were inoculated with *Pseudomonas putida* H-2-3 drought and salinity stress were alleviated as the organism had gibberellins secreting ability (Kang et al. 2014b).

Cytokinin stimulate division and enlargement of cell, elongation of shoot, uptake of nutrients, and prevention of senescence, vascular development and causes stomatal opening. *Platycladeus orientalis* container seedlings when inoculated with *Bacillus subtilis* mitigated drought stress due to the ability of bacteria to produce cytokinin, while the elevated levels of ABA reversed the potential damage caused to *Platycladeus orientalis* seedlings which was caused by increased cytokinin on stomatal opening. Under drought stress and salinity stress conditions, endogenous ABA is produced rapidly in response to water deficit by closing stomata to minimise water loss through transpiration. Additionally, it encourages root branching to enhance water absorption and stimulate leaf growth (Tardieu et al. 2010). Increase in the ABA and Jasmonic acid contents by *Sphingomonas* sp. LK11 leads to the improvement of plant biomass, photo-synthetic pigments, production of glycine, glutamate, and proline contents under drought stress in *Glycine max* (Asaf et al. 2017).

Osmolyte Dynamics and Metabolic Regulation for Drought Stress Relief Another factor affecting plant adaptation to drought is the accumulation of secondary metabolites in the cytoplasm. Plants primarily respond to drought stress through metabolites, often termed as osmotic adaptations. They serve as compatible solutes, aiding in the maintenance of plant metabolic activity (Kaur et al. 2017). During drought stress, these osmotic adjustment compounds accumulate in the cytoplasm, lowering the cell's water potential and maintaining turgor, this promotes the plant's uptake of water from the soil without interfering with normal metabolism. When plants are under drought stress, their metabolic profiles changes, which may indicate these changes cause signal transduction and the initiation of stress tolerance in plants. Additionally, these metabolites function as antioxidants and/or are part of plant's defence system (Kaplan et al. 2004; Anjum et al. 2011). As free radical scavengers, these osmo-protectants increases the antioxidant enzyme activity, prevent oxidation by removing extra ROS, and restore cellular redox equilibrium. Under stressful conditions plants accumulate osmolytes to escape from stress. These osmolytes include amino acids, proline, polyamines, organic acids, glucose, trehalose, maltose, glutamate, cellobiose, glycine betaine, polyhydric alcohols, and water stress proteins that includes dehydrins (Zhu et al. 2002; Couée et al. 2006).

Antioxidant Defence System Increase in drought stress leads into photorespiration, interruption of photosynthesis and cellular homeostasis. These results in the release of reactive oxygen species (ROS) (Farooq et al. 2009; Anjum et al. 2011; Kaur et al. 2017). ROS include singlet oxygen ($^1O_2$), superoxide anion radicals ($O_2^{\bullet-}$), hydroxyl radicals ($^{\bullet}OH$), hydrogen peroxide (H_2O_2), hydroxide (OH^-), per hydroxyl radical ($HO_2^{\bullet}$), and free and alkoxy radicals (RO) (Vurukonda et al. 2016). Under normal physiological condition, ROS are produced at low concentrations in a different cell organelle including peroxisomes, mitochondria, and chloroplasts. ROS are toxic and extremely reactive, harm DNA, proteins, lipids, carbohydrates, impairing normal plant cell activity and ultimately leading into programmed cell death (PCD).

PGPR aid plants to overcome oxidative stress by amplifying the work of antioxidant enzymes like catalase, superoxide dismutase and peroxides. They also synthesize secondary metabolites which function as antioxidants and elicit induced systemic resistance (ISR) (Chandran et al. 2021). Besides this they also produce osmo-protectants like glycine betaine and proline, enhance uptake of nutrient, modulate plant hormones under stress conditions and protect the root structure through synthesis of exopolysaccharides (EPS) (Figueiredo et al. 2008). Furthermore, PGPR could quench ROS directly thus protecting plants from oxidative damage and maintaining their healthiness and recoverability (Bharti et al. 2019).

Role of the Volatile Organic Compounds (VOCs) VOCs of plants are divided into heavier and smaller groups. Methanol and ethylene (ET) are examples of small VOCs, whereas methyl jasmonate, methyl salicylate and terpenes are examples of heavier VOCs. In general, the degree of stress is correlated with the amount of the volatile chemicals emitted. Stress-induced VOCs have been linked to plant growth enhancement and systemic defensive responses, acting as priming signalling agents (German et al. 2000; Copolovici et al. 2011). Furthermore, when plants sense signals from nearby plants that are experiencing drought, they activate their stress response mechanisms.

Elicitation of VOC emissions represent a quick promising approach because the efficacy of PGPR will be assessed to evaluate plants that are affected by drought (Timmusk et al. 2014). PGPR produces various highly diffusible VOCs into the soil and plant, including alcohols, ketones, and alkenes. Specific group of VOCs are produced by specific strain of bacteria, and these compounds are important in the life cycle of bacteria, directly impact development of plant, and indirectly regulate plant-microbial interactions (Peñuelas et al. 2014). Certain VOCs function as source of nutrition or as signalling molecules in plants (Bailly et al. 2014). VOCs stimulate seed germination and enhancing fruit production. Furthermore, the chlorophyll content of leaves and photosynthetic efficiency are increased which boost photosynthesis and increased production of carbohydrates. VOCs also promote nutrient uptake of iron and sulphur (Castulo-Rubio et al. 2015). Table 7.2 summarises different mechanisms used by PGPR to alleviate drought stress.

5 Conclusion

Rhizobacterial-plant interactions play a crucial role in enhancing plant resilience to drought and salinity stress, ultimately supporting plant growth and increasing crop yields under challenging environmental conditions. Significant advancements have been made in understanding how rhizobacteria promote drought and salt tolerance, with various studies highlighting their role in modulating plant physiology, enhancing nutrient uptake, and inducing stress-responsive pathways. However, achieving consistent and sustainable benefits across diverse agricultural systems requires a deeper understanding of the complex relationships among the plant microbiome, plant genotype, and environmental factors. Future research should focus on dissecting these intricate interactions to fully harness the potential of rhizobacteria in promoting abiotic stress tolerance. Such insights will be instrumental in developing targeted, microbiome-based strategies that improve plant survival, productivity, and resilience, contributing to more sustainable agricultural practices and food security.

References

Abulfaraj AA, Jalal RS (2021) Use of plant growth promoting bacteria to enhance salinity stress in soybean (*Glycine max* l.) plants. Saudi J Biol Sci 28:3823–3834. https://doi.org/10.1016/j.sjbs.2021.03.053

Acosta-Motos JR, Ortuño MF, Bernal-Vicente A, Diaz-Vivancos P, Sanchez-Blanco MJ, Hernandez JA (2017) Plant responses to salt stress: adaptive mechanisms. Agronomy 7(1):18

Ali B, Wang X, Saleem MH, Sumaira, Hafeez A, Afridi MS (2022) PGPR mediated salt tolerance in maize by modulating plant physiology, antioxidant defence, compatible solutes accumulation and bio-surfactant producing genes. Plants 11(3):345. https://doi.org/10.3390/plants11030345

Allotey DFK, Asiamah RD, Dedzoe CD, Nyamekye AL (2008) Physico-chemical properties of three salt-affected soils in the lower volta basin and management strategies for their sustainable utilization. West Afr J Appl Ecol 12:1–14. https://doi.org/10.4314/wajae.v12i1.45776

Amna S, Sarfraz B, Din Y, Xia MA, Kamran MT, Javed TS, Chaudhary HJ (2019) Mechanistic elucidation of germination potential and growth of wheat inoculated with exopolysaccharide and ACC-deaminase producing bacillus strains under induced salinity stress. Ecotoxicol Environ Saf 183:109466

Anjum SA, Xie XY, Wang LC, Saleem MF, Man C, Lei W (2011) Morphological, physiological and biochemical responses of plants to drought stress. Afr J Agric Res 6:2026–2032

Ansari F, Jabeen M, Ahmad I (2021) *Pseudomonas azotoformans* FAP5, a novel biofilm forming PGPR strain, alleviates drought stress in wheat plant. Int J Environ Sci Technol 18:1–16. https://doi.org/10.1007/s13762-020-03045-9

Ansari FA, Ahmad I, Pichtel J (2023) Synergistic effects of biofilm-producing PGPR strains on wheat plant colonization, growth and soil resilience under drought stress. Saudi J Biol Sci 30(6):103664

Apel K, Hirt H (2004) Reactive oxygen species: metabolism, oxidative stress, and signal transduction. Annu Rev Plant Biol 55:373–399. https://doi.org/10.1146/annurev.arplant.55.031903.141701. PMID: 15377225

Arora NK, Fatima T, Mishra J, Mishra I, Verma S, Verma R, Bharti C (2020) Halo-tolerant plant growth promoting rhizobacteria for improving productivity and remediation of saline soils. J Adv Res 26:69–82. https://doi.org/10.1016/j.jare.2020.07.003

Asaf S, Khan AL, Khan MA, Imran QM, Yun BW & Lee IJ (2017) Osmoprotective functions conferred to soybean plants via inoculation with *Sphingomonas* sp. LK11 and exogenous trehalose. Microbiol Res 205:135–145

Ayuso-Calles M, Flores-Félix JD, Rivas R (2021) Overview of the role of rhizobacteria in plant salt stress tolerance. Agronomy 11(9):1759. https://doi.org/10.3390/agronomy11091759

Azooz MM, Shaddad MA, Abdel-Latef AA (2004) Leaf growth and K+/Na+ ratio as an indication of the salt tolerance of three sorghum cultivars grown under salinity stress and IAA treatment. Acta Agron Hung 52(3):287–296

Backer R, Rokem JS, Ilangumaran G, Lamont J, Praslickova D, Ricci E (2018) Plant growth-promoting rhizobacteria: context, mechanisms of action, and roadmap to commercialization of bio-stimulants for sustainable agriculture. Front Plant Sci 9:1473

Bailly A, Groenhagen U, Schulz S, Geisler M, Eberl L, Weisskopf L (2014) The interkingdom volatile signal indole promotes root development by interfering with auxin signalling. Plant J 80(5):758–771

Bano A, Fatima M (2009) Salt tolerance in *Zea mays* (L). following inoculation with *Rhizobium* and *Pseudomonas*. Biol Fertil Soils 45:405–413. https://doi.org/10.1007/s00374-008-0344-9

Batool T, Ali S, Seleiman MF, Naveed NH, Ali A, Ahmed K, Mubushar M (2020) Plant growth promoting rhizobacteria alleviates drought stress in potato in response to suppressive oxidative stress and antioxidant enzymes activities. Sci Rep *10*(1):16975

Belimov AA, Dodd IC, Hontzeas N, Theobald JC, Safronova VI, Davies WJ (2009) Rhizosphere bacteria containing 1-aminocyclopropane-1-carboxylate deaminase increase yield of

plants grown in drying soil via both local and systemic hormone signalling. New Phytol *181*(2):413–423. https://doi.org/10.1111/j.1469-8137.2008.02657.x
Berendsen RL, Pieterse CM, Bakker PA (2012) The rhizosphere microbiome and plant health. Trends Plant Sci 17(8):478–486
Bharti N, Barnawal D (2019) Amelioration of salinity stress by PGPR: ACC deaminase and ROS scavenging enzymes activity. In: PGPR amelioration in sustainable agriculture. Woodhead Publishing, INDIA, belongs to Elsevier Publications, pp 85–106
Bharti N, Pandey SS, Barnawal D, Patel VK, Kalra A (2016) Plant growth promoting rhizobacteria *Dietzia natronolimnaea* modulates the expression of stress responsive genes providing protection of wheat from salinity stress. Sci Rep 6:34768. https://doi.org/10.1038/srep34768
Bhat MA, V, Bhat MA, Wani IA, Dar FL, Farooq I, Jan AT (2020) Mechanistic insights of the interaction of plant growth-promoting rhizobacteria (PGPR) with plant roots toward enhancing plant productivity by alleviating salinity stress. Front Microbiol 11:1952. https://doi.org/10.3389/fmicb.2020.01952
Bhattacharyya D, Yu SM, Lee YH (2015) Volatile compounds from *Alcaligenes faecalis* JBCS1294 confer salt tolerance in *Arabidopsis thaliana* through the auxin and gibberellin pathways and differential modulation of gene expression in root and shoot tissues. Plant Growth Regul 75:297–306. https://doi.org/10.1007/s10725-014-9953-5
Bomfeti CA, Florentino LA, Guimarães AP, Cardoso G, Guerreiro P (2011) Exopolysaccharides produced by the symbiotic nitrogen-fixing bacteria of Leguminosae. Rev Bras Ciênc Solo 35:657–671
Borlee BR, Goldman AD, Murakami K, Samudrala R, Wozniak DJ, Parsek MR (2010) *Pseudomonas aeruginosa* uses a cyclic-di-GMP-regulated adhesin to reinforce the biofilm extracellular matrix. Mol Microbiol 75(4):827–842
Camaille M, Fabre N, Clément C, Ait Barka E (2021) Advances in wheat physiology in response to drought and the role of plant growth promoting rhizobacteria to trigger drought tolerance. Microorganisms *9*(4):687. https://doi.org/10.3390/microorganisms9040687
Castillo P, Escalante M, Gallardo M, Alemano S, Abdala G (2013) Effects of bacterial single inoculation and co-inoculation on growth and phytohormone production of sunflower seedlings under water stress. Acta Physiol Plant 35:2299–2309
Castulo-Rubio DY, Alejandre-Ramírez NA, Orozco-Mosqueda MDC, Santoyo G, Macías-Rodríguez LI, Valencia-Cantero E (2015) Volatile organic compounds produced by the rhizobacterium *Arthrobacter agilis* UMCV2 modulate *Sorghum bicolor* (strategy II plant) morphogenesis and SbFRO1 transcription in vitro. J Plant Growth Regul 34:611–623
Chakdar H, Borse DN, Verma S, Choudhary P, Das S (2019) Microbial management of crop salinity stress: mechanisms, applications, and prospects. In: Akhtar MS (ed) Salt stress, microbes, and plant interactions: mechanisms and molecular approaches. Springer, Singapore, pp 1–25. https://doi.org/10.1007/978-981-13-8805-7_1
Chandra D, Srivastava R, Sharma AK (2018) Influence of IAA and ACC deaminase producing fluorescent pseudomonads in alleviating drought stress in wheat (*Triticum aestivum*). Agric Res 7:290–299
Chandra P, Singh A, Choudhary M, Yadav RK (2021) Role of plant growth promoting rhizobacteria (PGPR) in mitigating salt stress. In: Agriculturally important microorganisms. CRC Press, London, pp 65–90
Chandran H, Meena M, Swapnil P (2021) Plant growth-promoting rhizobacteria as a green alternative for sustainable agriculture. Sustainability 13:10986
Chatterjee P, Samaddar S, Niinemets Ü, Sa TM (2018) *Brevibacterium linens* RS16 confers salt tolerance to *Oryza sativa* genotypes by regulating antioxidant defence and H+ ATPase activity. Microbiol Res 215:89–101. https://doi.org/10.1016/j.micres.2018.06.007
Chaudhry S, Sidhu GPS (2021) Climate change regulated abiotic stress mechanisms in plants: a comprehensive review. Plant Cell Rep 41:1–31. https://doi.org/10.1007/s00299-021-02759-5

Chen L, Liu Y, Wu G, Veronican Njeri K, Shen Q, Zhang N, Zhang R (2016) Induced maize salt tolerance by rhizosphere inoculation of *Bacillus amyloliquefaciens* SQR9. Physiol Plant 158(1):34–44. https://doi.org/10.1111/ppl.12441

Choudhary M, Chandra P, Dixit B, Nehra V, Choudhary U, Chaudhry S (2022) Plant growth-promoting microbes: role and prospective in amelioration of salt stress. Commun Soil Sci Plant Anal 53(13):1692–1711. https://doi.org/10.1080/00103624.2022.2063316

Copolovici L, Kännaste A, Remmel T, Vislap V, Niinemets U (2011) Volatile emissions from *Alnus glutionosa* induced by herbivory are quantitatively related to the extent of damage. J Chem Ecol 37(1):18–28

Couée I, Sulmon C, Gouesbet G, El-Amrani A (2006) Involvement of soluble sugars in reactive oxygen species balance and responses to oxidative stress in plants. J Exp Bot 57(3):449–459

Curá JA, Franz DR, Filosofía JE, Balestrasse KB, Burgueño LE (2017) Inoculation with *Azospirillum* sp. and *Herbaspirillum* sp. bacteria increases the tolerance of maize to drought stress. Microorganisms 5(3):41

Dai A (2011) Drought under global warming: a review. Wires Clim Change 2:45–65

Danish S, Zafar-ul-Hye M (2019) Co-application of ACC-deaminase producing PGPR and timber-waste biochar improves pigments formation, growth and yield of wheat under drought stress. Sci Rep 9(1):5999.v

Danish S, Zafar-ul-Hye M, Fahad S (2020) Drought stress alleviation by ACC deaminase producing *Achromobacter xylosoxidans* and *Enterobacter cloacae*, with and without timber waste biochar in maize. Sustainability 12:6286

Dodd IC, Pérez-Alfocea F (2012) Microbial amelioration of crop salinity stress. J Exp Bot 63(9):3415–3428. https://doi.org/10.1093/jxb/ers033

Egamberdieva D (2012) *Pseudomonas chlororaphis*: a salt-tolerant bacterial inoculants for plant growth stimulation under saline soil conditions. Acta Physiol Plantarum 34:751–756

Egamberdieva D, Kucharova Z (2009) Selection for root colonising bacteria stimulating wheat growth in saline soils. Biol Fertil Soils 45:563–571

Etesami H (2018) Can interaction between silicon and plant growth promoting rhizobacteria benefit in alleviating abiotic and biotic stresses in crop plants? Agric Ecosyst Environ 253:98–112. https://doi.org/10.1016/j.agee.2017.11.007

Etesami H (2020) Plant microbe interactions in plants and stress tolerance. In: Tripathi DK (ed) *Plant life under changing environment*. Academic, Elsevier, London, pp 355–396. https://doi.org/10.1016/B978-0-12-818204-8.00018-7

Etesami H, Alikhani HA, Mirseyed Hosseini H (2015) Indole-3-acetic acid and 1-aminocycloprop ane-1-carboxylate deaminase: bacterial traits required in rhizosphere, rhizoplane and/or endophytic competence by beneficial bacteria. In: Bacterial metabolites in sustainable agroecosystem, pp 183–258. https://doi.org/10.1007/978-3-319-24654-3_8

Farooq M, Wahid A, Kobayashi N, Fujita D, Basra SMA (2009) Plant drought stress: effects, mechanisms and management. Agron Sustain Dev 29(1):185–212

Figueiredo MV, Burity HA, Martinez CR, Chanway CP (2008) Alleviation of drought stress in the common bean (*Phaseolus vulgaris* L.) by co-inoculation with *Paenibacillus polymyxa* and *Rhizobium tropici*. Appl Soil Ecol 40(1):182–188

Fischer SE, Miguel MJ, Mori GB (2003) Effect of root exudates on the exopolysaccharide composition and the lipopolysaccharide profile of *Azospirillum brasilense* cd under saline stress. FEMS Microbiol Lett 219(1):53–62

Forni C, Duca D, Glick BR (2017) Mechanisms of plant response to salt and drought stress and their alteration by rhizobacteria. Plant Soil 410:335–356. https://doi.org/10.1007/s11104-016-3007-x

Gamalero E, Glick BR (2022) Recent advances in bacterial amelioration of plant drought and salt stress. Biology 11(3):437. https://doi.org/10.3390/biology11030437

German MA, Burdman S, Okon Y, Kigel J (2000) Effects of *Azospirillum brasilense* on root morphology of common bean (*Phaseolus vulgaris* L.) under different water regimes. Biol Fertil Soils 32:259–264

Gill SS, Tuteja N (2010) Polyamines and abiotic stress tolerance in plants. Plant Signal Behav 5(1):26–33. https://doi.org/10.4161/psb.5.1.10291

Gill SS, Gill R, Trivedi DK, Anjum NA, Sharma KK, Ansari MW, Tuteja N (2016) *Piriformospora indica*: potential and significance in plant stress tolerance. Front Microbiol 7:332. https://doi.org/10.3389/fmicb.2016.00332

Ghosh D, Gupta A, Mohapatra S (2019) A comparative analysis of exopolysaccharide and phytohormone secretions by four drought-tolerant rhizobacterial strains and their impact on osmotic-stress mitigation in Arabidopsis thaliana. World J Microbiol Biotechnol 35(6):90

Glick BR (2014) Bacteria with ACC deaminase can promote plant growth and help to feed the world. Microbiol Res 169(1):30–39. https://doi.org/10.1016/j.micres.2013.09.009

Gopalakrishnan S, Humayun P, Srinivas S, Vijayabharathi R, Ratnakumari B, Rupela O (2012a) Plant growth-promoting traits of biocontrol potential *Streptomyces* isolated from herbal vermicompost. Biocontrol Sci Tech 22(10):1199–1210. https://doi.org/10.1080/09583157.2012.719151

Gopalakrishnan S, Upadhyaya H, Vadlamudi S, Humayun P, Vidya MS, Alekhya G, Singh A, Vijayabharathi R, Bhimineni RK, Seema M, Rathore A, Rupela O (2012b) Plant growth-promoting traits of biocontrol potential bacteria isolated from rice rhizosphere. Springerplus 1(1):71. https://doi.org/10.1186/2193-1801-1-71

Gopalakrishnan S, Srinivas V, Shravya A, Prakash B, Ratnakumari B, Vijayabharathi R, Rupela O (2013a) Evaluation of *Streptomyces* spp. for their plant growth-promoting traits in rice. Can J Microbiol 59:534–539. https://doi.org/10.1139/cjm-2013-0287

Gopalakrishnan S, Srinivas V, Sree Vidya M, Rathore A (2013b) Plant growth-promoting activities of *Streptomyces* spp. in sorghum and rice. Springerplus 2:574. https://doi.org/10.1186/2193-1801-2-574

Gopalakrishnan S, Srinivas V, Prakash B, Arumugam S, Vijayabharathi R, Rupela O, Kudapa H, Krishnamohan K, Varshney RK (2014) Evaluation of *Streptomyces* strains isolated from herbal vermicompost for their plant growth-promotion traits in rice. Microbiol Res 169:40–48. https://doi.org/10.1016/j.micres.2013.09.008

Gopalakrishnan S, Sathya A, Vijayabharathi R, Varshney RK, Gowda CLL, Krishnamurthy L (2015) Plant growth promoting rhizobia: challenges and opportunities. 3 Biotech 5:355–377

Gopalakrishnan S, Srinivas V, Alekhya G, Prakash B, Kudapa H, Varshney RK (2015a) Evaluation of *Streptomyces* sp. obtained from herbal vermicompost for broad spectrum of plant growth-promoting activities in chickpea. Org Agric 5:123–133. https://doi.org/10.1007/s13165-015-0099-1

Gopalakrishnan S, Srinivas V, Alekhya G, Prakash B, Kudapa H, Varshney RK (2015b) Evaluation of broad-spectrum *Streptomyces* sp. for plant growth promotion traits in chickpea (*Cicer arietinum* L.). Phil Agric Sci 98(3):270–278. http://oar.icrisat.org/id/eprint/9056

Gopalakrishnan S, Srinivas V, Alekhya G, Prakash B (2015c) Effect of plant growth-promoting *Streptomyces* sp. on growth promotion and grain yield in chickpea (*Cicer arietinum* L). 3 Biotech 5:799–806. https://doi.org/10.1007/s13205-015-0283-8

Gopalakrishnan S, Sathya A, Vijayabharathi R (2016a) Plant growth-promoting actinobacteria: a new avenue for enhancing the productivity & soil fertility of grain legumes. Springer Singapore. isbn:978-981-10-0705-7

Gopalakrishnan S, Srinivas V, Sameer CV (2016b) Plant growth-promotion traits of *Streptomyces* sp. in pigeonpea. Legume Perspect 11:43–44. http://oar.icrisat.org/id/eprint/9238

Gopalakrishnan S, Vadlamudi S, Samineni S, Sameer CV (2016c) Plant growth-promotion and biofortification of chickpea and pigeonpea through inoculation of biocontrol potential bacteria, isolated from organic soils. Springerplus 5(1):1882. https://doi.org/10.1186/s40064-016-3590-6

Gopalakrishnan S, Srinivas V, Chand U, Pratyusha S, Samineni S (2022) *Streptomyces* consortia-mediated plant growth-promotion and yield performance in chickpea. 3 Biotech 12(11):318. https://doi.org/10.1007/s13205-022-03389-8

Groppa MD, Benavides MP, Zawoznik MS (2012) Root hydraulic conductance, aquaporins and plant growth promoting microorganisms: a revision. Appl Soil Ecol 61:247–254. https://doi.org/10.1016/j.apsoil.2011.11.013

Gupta S, Pandey S (2019) ACC deaminase producing bacteria with multifarious plant growth promoting traits alleviates salinity stress in French bean (*Phaseolus vulgaris*) plants. Front Microbiol 10:1506

Gururani MA, Upadhyaya CP, Baskar V, Venkatesh J, Nookaraju A, Park SW (2013) Plant growth-promoting rhizobacteria enhance abiotic stress tolerance in *Solanum tuberosum* through inducing changes in the expression of ROS-scavenging enzymes and improved photosynthetic performance. J Plant Growth Regul 32:245–258

Hamdia MAES, Shaddad MAK, Doaa MM (2004) Mechanisms of salt tolerance and interactive effects of *Azospirillum brasilense* inoculation on maize cultivars grown under salt stress conditions. Plant Growth Regul 44:165–174. https://doi.org/10.1023/B:GROW.0000049414.03099.9b

Haque MM, Mosharaf MK, Khatun M, Haque MA, Biswas MS, Islam MS, Siddiquee MA (2020) Biofilm producing rhizobacteria with multiple plant growth-promoting traits promote growth of tomato under water-deficit stress. Front Microbiol 11:542053

He Z, He C, Zhang Z, Zou Z, Wang H (2007) Changes of anti-oxidative enzymes and cell membrane osmosis in tomato colonized by arbuscular mycorrhizae under NaCl stress. Colloids Surf B Biointerfaces 59(2):128–133. https://doi.org/10.1016/j.colsurfb.2007.04.023

Hou C, Li X, Tian D, Xu B, Zhang C, Ren J, Chen N (2022) Evaluation of the effects of water and salinity stress on the growth and biochemistry of alfalfa (*Medicago sativa* L.) at the branching stage. Sustainability 14(16):10262

Ilangumaran G, Smith DL (2017) Plant growth promoting rhizobacteria in amelioration of salinity stress: a systems biology perspective. Front Plant Sci 8:275185. https://doi.org/10.3389/fpls.2017.01768

Ings J, Mur LA, Robson PR, Bosch M (2013) Physiological and growth responses to water deficit in the bioenergy crop *Miscanthus x giganteus*. Front Plant Sci 4:468

Kang SM, Khan AL, Waqas M, You YH, Kim JH, Kim JG, Lee IJ (2014a) Plant growth-promoting rhizobacteria reduce adverse effects of salinity and osmotic stress by regulating phytohormones and antioxidants in *Cucumis sativus*. J Plant Interact 9(1):673–682

Kang SM, Radhakrishnan R, Khan AL, Kim MJ, Park JM, Kim BR, Lee IJ (2014b) Gibberellin secreting rhizobacterium, *Pseudomonas putida* H-2-3 modulates the hormonal and stress physiology of soybean to improve the plant growth under saline and drought conditions. Plant Physiol Biochem 84:115–124

Kang SM, Shahzad R, Bilal S, Khan AL, Park YG, Lee KE, Lee IJ (2019) Indole-3-acetic-acid and ACC deaminase producing *Leclercia adecarboxylata* MO1 improves *Solanum lycopersicum* L. growth and salinity stress tolerance by endogenous secondary metabolites regulation. BMC Microbiol 19:1–14

Kaplan F, Kopka J, Haskell DW, Zhao W, Schiller KC, Gatzke N (2004) Exploring the temperature-stress metabolome of *Arabidopsis*. Plant Physiol 136(4):4159–4168

Kasotia A, Varma A, Tuteja N, Choudhary DK (2016) Amelioration of soybean plant from saline-induced condition by exopolysaccharide producing *Pseudomonas*-mediated expression of high affinity K+-transporter (HKT1) gene. Curr Sci 111(12):1961–1967

Kaur G, Asthir B (2017) Molecular responses to drought stress in plants. Biol Plant 61:201–209

Kaushal M, Wani SP (2016) Rhizobacterial-plant interactions: strategies ensuring plant growth promotion under drought and salinity stress. *Agric Ecosyst Environ* 231:68–78. https://doi.org/10.1016/j.agee.2016.06.031

Khan MA, Iqbal M, Akram M, Ahmad M, Hassan MW, Jamil M (2013) Recent advances in molecular tool development for drought tolerance breeding in cereal crops: a review. Zemdirb Agric 100:325–334

Khan MA, Asaf S, Khan AL, Ullah I, Ali S, Kang SM, Lee IJ (2019a) Alleviation of salt stress response in soybean plants with the endophytic bacterial isolate *Curtobacterium* sp. SAK1. Ann Microbiol 69:797–808. https://doi.org/10.1007/s13213-019-01470-x

Khan MA, Asaf S, Khan AL, Jan R, Kang SM, Kim KM, Lee IJ (2019b) Rhizobacteria AK1 remediates the toxic effects of salinity stress via regulation of endogenous phytohormones and gene expression in soybean. Biochem J 476(16):2393–2409. https://doi.org/10.1042/BCJ20190435

Khan A, Bilal S, Khan AL, Imran M, Al-Harrasi A, Al-Rawahi A, Lee IJ (2020a) Silicon-mediated alleviation of combined salinity and cadmium stress in date palm (*Phoenix dactylifera L.*) by regulating physio-hormonal alteration. EES 188:109885

Khan MA, Asaf S, Khan AL, Adhikari A, Jan R, Ali S, Lee IJ (2020b) Plant growth-promoting endophytic bacteria augment growth and salinity tolerance in rice plants. Plant Biol 22(5):850–862

Kim K, Jang YJ, Lee SM, Oh BT, Chae JC, Lee KJ (2014) Alleviation of salt stress by Enterobacter sp. EJ01 in tomato and *Arabidopsis* is accompanied by up-regulation of conserved salinity responsive factors in plants. Mol Cells 37(2):109–117. https://doi.org/10.14348/molcells.2014.2239

A, Patel J, Meena VS, Ramteke P (2019a) Plant growth promoting rhizobacteria: strategies to improve abiotic stresses under sustainable agriculture. J Plant Nutr 42:1402–1415

A, Patel JS, Meena VS, Srivastava R (2019b) Recent advances of PGPR based approaches for stress tolerance in plants for sustainable agriculture. Biocatal Agric Biotechnol 20:101271. https://doi.org/10.1016/j.bcab.2019.101271

A, Singh S, Gaurav AK, Srivastava S, Verma JP (2020) Plant growth-promoting bacteria: biological tools for the mitigation of salinity stress in plants. Front Microbiol 11:1216. https://doi.org/10.3389/fmicb.2020.01216

V, Raghuvanshi N, Pandey AK, A, Thoday-Kennedy E, Kant S (2023) Role of halotolerant plant growth-promoting rhizobacteria in mitigating salinity stress: recent advances and possibilities. Agriculture *13*(1):168

Kumawat KC, Nagpal S, Sharma P (2022) Potential of plant growth-promoting rhizobacteria-plant interactions in mitigating salt stress for sustainable agriculture: a review. Pedosphere 32(2):223–245. https://doi.org/10.1016/S1002-0160(21)60070-X

Lerner A, Castro-Sowinski S, Valverde A, Lerner H, Dror R, Okon Y, Burdman S (2009) The *Azospirillum brasilense* Sp7 noeJ and noeL genes are involved in extracellular polysaccharide biosynthesis. Microbiology 155(12):4058–4068

Li Q, Dong F, Dai Q, Zhang C, Yu L (2018) Surface properties of PM2.5 calcite fine particulate matter in the presence of same-size bacterial cells and exocellular polymeric substances (EPS) of *Bacillus mucitaginosus*. Environ Sci Pollut Res 25(23):22429–22436

Li H, Qiu Y, Yao T, Ma Y, Zhang H, Yang X (2020) Effects of PGPR microbial inoculants on the growth and soil properties of *Avena sativa*, *Medicago sativa,* and *Cucumis sativus* seedlings. Soil Tillage Res 199:104577. https://doi.org/10.1016/j.still.2020.104577

Liu X, Luo Y, Li Z, Wang J, and Wei G (2017) Role of exopolysaccharide in salt stress resistance and cell motility of *Mesorhizobium alhagi* CCNWXJ12–2T. Appl Microbiol Biotechnol 101(7):2967–29786

Liu J, Shabala S, Zhang J, Ma G, Chen D, Shabala L, Zhao Q (2020) Melatonin improves rice salinity stress tolerance by NADPH oxidase-dependent control of the plasma membrane K+ transporters and K+ homeostasis. Plant Cell Environ 43(11):2591–2605

Liu X, Chai J, Zhang Y, Zhang C, Lei Y, Li Q, Yao T (2022) Halotolerant rhizobacteria mitigate the effects of salinity stress on maize growth by secreting exopolysaccharides. Environ Exp Bot 204:105098. https://doi.org/10.1016/j.envexpbot.2022.105098

Ma Y, Zhang W, Niu J, Ren Y, Zhang F (2018) Hydrogen sulfide may function downstream of hydrogen peroxide in salt stress induced stomatal closure in *Vicia faba*. Funct Plant Biol 46(2):136–145. https://doi.org/10.1071/FP18096

Mahmud-Ur-Rahman Naser IB, Mahmud NU, Sarker A, Hoque MN, Islam TA (2022) Highly salt-tolerant bacterium *Brevibacterium sediminis* promotes the growth of rice (*Oryza sativa* L.) seedlings. Stress 2:275–289. https://doi.org/10.3390/stresses2030020

Martínez C, Espinosa-Ruiz A, Prat S (2016) Gibberellins and plant vegetative growth. Annu Plant Rev 49:285–322. https://doi.org/10.1002/9781119210436.ch10

Mauchline TH, Malone JG (2017) Life in earth–the root microbiome to the rescue? Curr Opin Microbiol 37:23–28

Mellidou I, Koukounaras A, Kostas S, Patelou E, Kanellis AK (2021) Regulation of vitamin C accumulation for improved tomato fruit quality and alleviation of abiotic stress. Genes 12:694

Misra S, Chauhan PS (2020) ACC deaminase-producing rhizosphere competent Bacillus spp. mitigate salt stress and promote *Zea mays* growth by modulating ethylene metabolism. 3 Biotech 10:119

Morel MA, Castro-Sowinski S (2013) The complex molecular signalling network in microbe–plant interaction. In: Plant microbe symbiosis: fundamentals and advances. Springer India, New Delhi, pp 169–199

Mukhopadhyay R, Sarkar B, Jat HS, Sharma PC, Bolan NS (2021) Soil salinity under climate change: challenges for sustainable agriculture and food security. J Environ Manag 280:111736. https://doi.org/10.1016/j.jenvman.2020.111736

Munns R, Tester M (2008) Mechanisms of salinity tolerance. Annu Rev Plant Biol 59(1):651–681. https://doi.org/10.1146/annurev.arplant.59.032607.092911

Naamala J, Smith DL (2021) Microbial derived compounds are a promising approach to mitigating salinity stress in agricultural crops. Front Microbiol 12:765320. https://doi.org/10.3389/fmicb.2021.765320

Nadeem SM, Zahir ZA, Naveed M, Arshad M (2009) Rhizobacteria containing ACC-deaminase confer salt tolerance in maize grown on salt-affected fields. Can J Microbiol 55(11):1302–1309. https://doi.org/10.1139/W09-092

Nadeem SM, Ahmad M, Zahir ZA, Javaid A, Ashraf M (2014) The role of mycorrhizae and plant growth promoting rhizobacteria (PGPR) in improving crop productivity under stressful environments. Biotechnol Adv 32(2):429–448

Nagpal S, Sharma P, Kumawat KC (2019) Assessment of native single and dual inoculants of *Mesorhizobium* sp. and endophytic rhizobacteria for plant growth promotion in chickpea. Agric Res J 56(4):746–751

Nascimento FX, Hernández AG, Glick BR, Rossi MJ (2020) Plant growth promoting activities and genomic analysis of the stress resistant *Bacillus megaterium* STB1, a bacterium of agricultural and biotechnological interest. Biotechnol Rep 25:e00406. https://doi.org/10.1016/j.btre.2019.e00406

Nascimento FX, Urón P, Glick BR, Giachini A, Rossi MJ (2021) Genomic analysis of the 1-amino cyclopropane-1-carboxylate deaminase-producing *Pseudomonas thivervalensis* sc5 reveals its multifaceted roles in soil and in beneficial interactions with plants. Front Microbiol 12:752288

Nian L, Zhang X, Yi X, Liu X, Ain NU, Yang Y, Zhu X (2021) Genome-wide identification of ABA receptor PYL/RCAR gene family and their response to cold stress in *Medicago sativa* L. Physiol Mol Biol Plants 27:1979–1995. https://doi.org/10.1007/s12298-021-01066-3

Niu SQ, Li HR, Pare PW, Aziz M, Wang SM, Shi HZ (2016) Induced growth promotion and higher salt tolerance in the halophyte grass *Puccinellia tenuiflora* by beneficial rhizobacteria. Plant Soil 407:217–230. https://doi.org/10.1007/s11104-015-2767-z

Niu X, Song L, Xiao Y, Ge W (2018) Drought-tolerant plant growth-promoting rhizobacteria associated with foxtail millet in a semi-arid agroecosystem and their potential in alleviating drought stress. Front Microbiol 8:2580

Noori Z, Delavar MA, Safari Y, Alavi-Siney SM (2021) Reclamation of a calcareous sodic soil with combined amendments: interactive effects of chemical and organic materials on soil chemical properties. Arab J Geosci 14:1–11. https://doi.org/10.1007/s12517-021-06485-w

Oleńska E, Małek W, Wójcik M, Swiecicka I, Thijs S, Vangronsveld J (2020) Beneficial features of plant growth-promoting rhizobacteria for improving plant growth and health in challenging conditions: a methodical review. Sci Total Environ 743:140682. https://doi.org/10.1016/j.scitotenv.2020.140682

Otlewska A, Migliore M, Dybka-Stępień K, Manfredini A, Struszczyk-Świta K, Napoli R, inzari F (2020) When salt meddles between plant, soil, and microorganisms. Front Plant Sci 11:553087. https://doi.org/10.3389/fpls.2020.553087

Patel D, Jha CK, Tank N, Saraf M (2012) Growth enhancement of chickpea in saline soils using plant growth promoting rhizobacteria. J Plant Growth Regul 31(1):53–62. https://doi.org/10.1007/s00344-011-9219-7

Paul D, Lade H (2014) Plant-growth-promoting rhizobacteria to improve crop growth in saline soils: a review. Agron Sustain Dev 34:737–752. https://doi.org/10.1007/s13593-014-0233-6

Peng J, Ma J, Wei X, Zhang C, Jia N, Wang X (2021) Accumulation of beneficial bacteria in the rhizosphere of maize (*Zea mays* L.) grown in a saline soil in responding to a consortium of plant growth promoting rhizobacteria. Ann Microbiol 71:40. https://doi.org/10.1186/s13213-021-01650-8

Peñuelas J, Asensio D, Tholl D, Wenke K, Rosenkranz M, Piechulla B (2014) Biogenic volatile emissions from the soil. Plant Cell Environ 37(8):1866–1891

Pereira SIA, Abreu D, Moreira H, Vega A, Castro PML (2020) Plant growth-promoting rhizobacteria (PGPR) improve the growth and nutrient use efficiency in maize (*Zea mays* L.) under water deficit conditions. Heliyon 6(10), volume 6, Issue 10, e05106, ISSN 2405-8440

Pereyra MA, Garcia P, Colabelli MN, Barassi CA, Creus CM (2012) A better water status in wheat seedlings induced by Azospirillum under osmotic stress is related to morphological changes in xylem vessels of the coleoptile. Appl Soil Ecol 53:94–97

Perret X, Staehelin C, Broughton WJ (2000) Molecular basis of symbiotic promiscuity. Microbiol Mol Biol Rev 64(1):180–201

Petrić I, Šamec D, Karalija E, Salopek-Sondi B (2022) Beneficial microbes and molecules for mitigation of soil salinity in Brassica species: a review soil systems. Soil Syst 6(1):18

Potopova V, Boroneanţ C, Boincean B, Soukup J (2016) Impact of agricultural drought on main crop yields in the Republic of Moldova. Int J Climatol 36:2063–2082

Prajapati M, Panchal R, Krupa Prajapati S (2022) Rhizobacterial-plant interaction approaches that enhance plant growth under abiotic stress. Acta Sci Agric. (ISSN: 2581-365X) 6(5). https://doi.org/10.31080/ASAG.2022.06.1136

Qiu QS, Guo Y, Dietrich MA, Schumaker KS, Zhu JK (2002) Regulation of SOS1, a plasma membrane Na^+/H^+ exchanger in *Arabidopsis thaliana*, by SOS2 and SOS3. Proc Natl Acad Sci 99(12):8436–8441. https://doi.org/10.1073/pnas.122224699

Qurashi AW, Sabri AN (2012) Bacterial exopolysaccharide and biofilm formation stimulate chickpea growth and soil aggregation under salt stress. Braz J Microbiol 43:1183–1191

Rashid U, Yasmin H, Hassan MN, Naz R, Nosheen A, Sajjad M, Ahmad P (2022) Drought-tolerant *Bacillus megaterium* isolated from semi-arid conditions induces systemic tolerance of wheat under drought conditions. Plant Cell Rep: 1–21, 41(3):549–569. https://doi.org/10.1007/s00299-020-02640-x. Epub 2021 Jan 7. PMID: 33410927

Rojas-Tapias D, Moreno-Galván A, Pardo-Díaz S, Obando M, Rivera D, Bonilla R (2012) Effect of inoculation with plant growth-promoting bacteria (PGPB) on amelioration of saline stress in maize (*Zea mays*). Appl Soil Ecol 61:264–272. https://doi.org/10.1016/j.apsoil.2012.01.006

Ruppel S, Franken P, Witzel K (2013) Properties of the halophyte microbiome and their implications for plant salt tolerance. Funct Plant Biol 40(9):940–951

Safdar H, Amin A, Shafiq Y, Ali A, Yasin R, Shoukat A, Sarwar MI (2019) A review: impact of salinity on plant growth. Nat Sci 17(1):34–40. https://doi.org/10.7537/marsnsj170119.06

Safdarian M, Askari H, Shariati JV, Nematzadeh G (2019) Transcriptional responses of wheat roots inoculated with *Arthrobacter nitroguajacolicus* to salt stress. Sci Rep 9(1):1792

Sagar A, Sayyed RZ, Ramteke PW, Sharma S, Marraiki N, Elgorban AM (2020) ACC deaminase and antioxidant enzymes producing halophilic *Enterobacter* sp. PR14 promotes the growth of rice and millets under salinity stress. Physiol Mol Biol Plants 26(9):1847–1854. https://doi.org/10.1007/s12298-020-00852-9

Saikia J, Sarma RK, Dhandia R, Yadav A, Bharali R, Gupta VK, Saikia R (2018) Alleviation of drought stress in pulse crops with ACC deaminase producing rhizobacteria isolated from acidic soil of Northeast India. Sci Rep 8(1):3560

Sánchez-Bermúdez M, Del Pozo JC, Pernas M (2022) Effects of combined abiotic stresses related to climate change on root growth in crops. Front Plant Sci 13:918537. https://doi.org/10.3389/fpls.2022.918537

Sandhya V, Ali SZ, Grover M, Reddy G, Venkateswarlu B (2009) Alleviation of drought stress effects in sunflower seedlings by the exopolysaccharides producing *Pseudomonas putida* strain GAP-P45. Biol Fertil Soils 46(1):17–26

Sapre S, Gontia-Mishra I, Tiwari S (2021) Plant growth promoting rhizobacteria ameliorates salinity stress in pea (*Pisum sativum*). J Plant Growth Regul 41:647–656. https://doi.org/10.1007/s00344-021-10329-y

Shah G, Jan M, Afreen M, Anees M, Rehman S, Daud MK, Malook I, Jamil M (2017) Halophilic bacteria mediated phytoremediation of salt-affected soils cultivated with rice. J Geochem Explor 174:59–65

Sharma S, Chandra D, Sharma AK (2021) Rhizosphere plant–microbe interactions under abiotic stress. In: Gupta VVSR, Sharma AK (eds) Rhizosphere biology: interactions between microbes and plants, rhizosphere biology. Springer, Singapore, pp 195–216. https://doi.org/10.1007/978-981-15-6125-2_10

Shilev S (2020) Plant-growth-promoting bacteria mitigating soil salinity stress in plants. Appl Sci 10(20):7326. https://doi.org/10.3390/app10207326

Shrivastava P, Kumar R (2015) Soil salinity: a serious environmental issue and plant growth promoting bacteria as one of the tools for its alleviation. Saudi J Biol Sci 22(2):123–131

Silva R, Filgueiras L, Santos B, Coelho M, Silva M, Estrada-Bonilla G, Meneses C (2020) *Gluconacetobacter diazotrophicus* changes the molecular mechanisms of root development in *Oryza sativa* L. growing under water stress. Int J Mol Sci 21(1):333

Singh DP, Singh V, Gupta VK, Shukla R, Prabha R, Sarma BK, Patel JS (2020) Microbial inoculation in rice regulates antioxidative reactions and defence related genes to mitigate drought stress. Sci Rep 10(1):4818

Singh RP, Ma Y, Shadan A (2022a) Perspective of ACC deaminase producing bacteria in stress agriculture. J Biotechnol 352:36–46. https://doi.org/10.1016/j.jbiotec.2022.05.002

Singh RP, Pandey DM, Jha PN, Ma Y (2022b) ACC deaminase producing rhizobacterium *Enterobacter cloacae* ZNP-4 enhance abiotic stress tolerance in wheat plant. PLoS One 17(5):e0267127. https://doi.org/10.1371/journal.pone.0267127

Song Y, Zheng H, Sui Y, Li S, Wu F, Sun X, Sui N (2022) SbWRKY55 regulates sorghum response to saline environment by its dual role in abscisic acid signalling. Theor Appl Genet 135(8):2609–2625. https://doi.org/10.1007/s00122-022-04130-y

Srinivas V, Gopalakrishnan S, Kamidi JP, Chander G (2020) Effect of plant growth-promoting *streptomyces* sp. on plant growth and yield of tomato and chilli. Andhra Pradesh J Agric Sci 6(2):65–70. http://oar.icrisat.org/id/eprint/11668.

Srinivas V, Naresh N, Pratyusha S, Ankati S, Govindaraj M, Gopalakrishnan S (2022) Enhancing pearl millet hybrids performance for yield and nutrients through plant growth-promoting *Streptomyces* spp.: an agronomic biofortification approach. Crop Pasture Sci 73(5):484–493. https://doi.org/10.1071/CP21438

Sultana S, Paul SC, Parveen S, Alam S, Rahman N, Jannat B, Hoque S, Rahman MT, Karim MM (2020) Isolation and identification of salt-tolerant plant-growth-promoting rhizobacteria and their application for rice cultivation under salt stress. Can J Microbiol 66(2):144–160. https://doi.org/10.1139/cjm-2019-0323. Epub 2019 Nov 12. PMID:31714812

Talei D, Yusop MK, Kadir MA, Valdiani A, Abdullah MP (2012) Response of King of Bitters (*Andrographis paniculata* Nees.) seedlings to salinity stress beyond the salt tolerance threshold. Aust J Crop Sci 6(6):1059–1067. https://search.informit.org/doi/10.3316/informit.734315006053753

Tardieu F, Parent B, Simonneau T (2010) Control of leaf growth by abscisic acid: hydraulic or non-hydraulic processes? Plant Cell Environ 33(4):636–647

Timmusk S, Abd El-Daim IA, Copolovici L, Tanilas T, Kännaste A, Behers L, Niinemets Ü (2014) Drought-tolerance of wheat improved by rhizosphere bacteria from harsh environments: enhanced biomass production and reduced emissions of stress volatiles. PLoS One 9(5):e96086

Upadhyay SK, Singh DP (2015) Effect of salt-tolerant plant growth promoting rhizobacteria on wheat plants and soil health in a saline environment. Plant Biol 17:288–293. https://doi.org/10.1111/plb.12173

Upadhyay SK, Singh JS, Singh DP (2011) Exopolysaccharide-producing plant growth-promoting rhizobacteria under salinity condition. Pedosphere 21(2):214–222

Vaishnav A, i S, Jain S, Varma A, Choudhary DK (2015) Putative bacterial volatile-mediated growth in soybean (*Glycine max* L. Merrill) and expression of induced proteins under salt stress. J Appl Microbiol 119(2):539–551. https://doi.org/10.1111/jam.12866

Vaishnav A, Varma A, Tuteja N, Choudhary DK (2016) PGPR-mediated amelioration of crops under salt stress. In: Plant-microbe interaction: an approach to sustainable agriculture, pp 205–226. https://doi.org/10.1007/978-981-10-2854-0_10

Vanbleu E, Choudhury BP, Carlson RW, Vanderleyden J (2005) The nodPQ genes in *Azospirillum brasilense* Sp7 are involved in sulfation of lipopolysaccharides. Environ Microbiol 7(11):1769–1774

Vigani G, Rolli E, Marasco R, Dell'Orto M, Michoud G, Soussi A, Daffonchio D (2019) Root bacterial endophytes confer drought resistance and enhance expression and activity of a vacuolar H+-pumping pyrophosphatase in pepper plants. Environ Microbiol 21(9):3212–3228

Vurukonda SSKP, Vardharajula S, Shrivastava M, SkZ A (2016) Enhancement of drought stress tolerance in crops by plant growth promoting rhizobacteria. Microbiol Res 184:13–24

Watts N, Amann M, Arnell N, Ayeb-Karlsson S, Beagley J, Belesova K (2021) The 2020 report of the lancet countdown on health and climate change: responding to converging crises. Lancet 397:129–170. https://doi.org/10.1016/S0140-6736(20)32290-X

Woo OG, Kim H, Kim JS, Keum HL, Lee KC, Sul WJ, Lee JH (2020) *Bacillus subtilis* strain GOT9 confers enhanced tolerance to drought and salt stresses in *Arabidopsis thaliana* and *Brassica campestris*. Plant Physiol Biochem 148:359–367

Wu M, Liu H, Han G, Cai R, Pan F, Xiang Y (2017) A moso bamboo WRKY gene PeWRKY83 confers salinity tolerance in transgenic Arabidopsis plants. Sci Rep 7:11721

Yadav AN, Yadav N (2018) Stress-adaptive microbes for plant growth promotion and alleviation of drought stress in plants. Acta Sci Agric 2:85–88

Yao L, Wu Z, Zheng Y, Kaleem I, Li C (2010) Growth promotion and protection against salt stress by *Pseudomonas putida* Rs-198 on cotton. Eur J Soil Biol 46:49–54. https://doi.org/10.1016/j.ejsobi.2009.11.002

Yu Y, Wang N, Hu R, Xiang F (2016) Genome-wide identification of soybean WRKY transcription factors in response to salt stress. Springerplus 5:920

Zhao C, Liu B, Piao S, Wang X, Lobell DB, Huang Y, Asseng S (2017) Temperature increase reduces global yields of major crops in four independent estimates. Proc Natl Acad Sci U S A 114(35):9326–9331

Zhao C, Zhang H, Song C, Zhu JK, Shabala S (2020) Mechanisms of plant responses and adaptation to soil salinity. Innovation (Camb) 1(1):100017. https://doi.org/10.1016/j.xinn.2020.100017

Zhu JK (2002) Salt and drought stress signal transduction in plants. Annu Rev Plant Biol 53:247–273

Chapter 8
Soil-Inhabiting Beneficial Fungi for Plant Growth-Promotion and Crop Protection Against Environmental Stresses in Iran

Parissa Taheri

1 Introduction

The increasing demand for organic agricultural products has incited interest in determining the role of various types of beneficial microorganisms in enhancing plant growth and health. Plant growth-promoting microorganisms (PGPMs), including numerous species of soil-inhabiting fungi, have been demonstrated as key players in promoting both plant and soil health via various cellular, molecular, biochemical and physiological mechanisms (Taheri 2022; Wei et al. 2024). These fungi contribute to nutrient recycling and solubilization, improve soil structure, and induce plant defense responses to biotic and abiotic environmental stimuli, which make them valuable in sustainable agriculture (Daroodi and Taheri 2024; Frac et al. 2018).

In the context of global challenges, such as climate change, soil degradation, and food security, exploring beneficial fungal species, understanding their modes of action in plant growth-promotion or defense activation, formulation and application of these fungi and their metabolites are crucial and can be more focused in future investigations. Numerous mechanisms might be involved in vital effects of beneficial soil fungi in agricultural systems, including biochemical cycling of nutrients, detoxification or bioremediation of soil pollutants, and direct or indirect effects on plant growth factors (Daroodi et al. 2022; Wilpiszeski et al. 2019).

Soil-inhabiting beneficial fungi play a crucial role in agriculture (Taheri and Flaherty 2022), particularly in Iran, where agricultural practices face numerous challenges such as soil degradation, water deficiency, salinity, together with other abiotic and biotic stresses. These fungi significantly improve plant growth and

P. Taheri (✉)
Department of Plant Protection, Faculty of Agriculture, Ferdowsi University of Mashhad, Mashhad, Iran
e-mail: p-taheri@um.ac.ir

S. Gnanamanickam, M. Höfte (eds.), *Disease Suppressive Tropical Soils*,
https://doi.org/10.1007/978-3-032-08100-1_8

development, making them essential for enhancing agricultural productivity and sustainability. One of the main functions of beneficial soil-inhabiting fungi is their ability to form symbiotic relationships with plant roots, which can be observed for the mycorrhizal and endophytic fungi. This symbiosis not only improves plant nutrition, but also increases plant resilience to environmental stresses such as drought and salinity, which are prevalent in many regions of Iran. By facilitating nutrient cycling and improving soil structure, these fungi contribute to healthier crops and more sustainable agricultural practices.

In addition to nutrient acquisition, beneficial fungi also play a vital role in disease suppression. They can be effective in controlling various phytopathogens in the soil, thereby reducing the severity and incidence of soil-borne diseases which threaten crop yields (Kandil et al. 2024; Wang et al. 2023; Yao et al. 2024). For instance, numerous strains of *Trichoderma* spp. (Badakhshan et al. 2022; Shahiri Tabarestani 2022), endophytic (Daroodi et al. 2022; Nooralvandi et al. 2021; Zare et al. 2023) and arbuscular mycorrhizal (AM) fungi (Mehdinejad et al. 2021) have been reported in Iran, so far. Some of these fungi are well-known for their plant growth-promotion and crop protection properties (Daroodi and Taheri 2024). Biocontrol or bioprotectant capability of these beneficial fungi is very important, because the reliance on chemical pesticides to control pathogens and insect pests can lead to environmental and health concerns.

Application of plant growth-promoting fungi (PGPF) in agriculture aligns with the global shift towards environmentally safe farming practices. Integrating these fungi into agricultural systems can reduce the need for chemical fertilizers and synthetic pesticides, leading to promoting eco-friendly approaches for plant production. The importance of soil-inhabiting PGPF contributes to the long-term sustainability of soil ecosystems. By enhancing soil fertility and structure, these fungi help in maintaining the ecological balance necessary for healthy agricultural systems. As Iran continues to face challenges related to food security and environmental sustainability, the role of beneficial fungi in agriculture seems to be increasingly important in this country.

2 Identification, Abundance and Diversity of Soil-Inhabiting Beneficial Fungi in Iran

Numerous beneficial endophytic, rhizospheric and mycorrhizal fungi resident in soil, belonging to Ascomycota, Basidiomycota, Glomeromycota, Mucoromycota, Chytridiomycota, and other phyla have been isolated and identified from Iran using morphological and molecular techniques. Conserved genomic regions play an important role in the molecular identification of various fungi, including soil-inhabiting fungal species. These regions are typically characterized by their stability across different species, making them ideal for phylogenetic analyses and species identification. The Internal Transcribed Spacer (ITS) regions of ribosomal RNA genes (ITS-rDNA) are among the most widely used markers for fungal identification. They provide sufficient variability for distinguishing between closely related species while maintaining conservation across broader taxonomic groups. High-quality fungal ITS sequences are essential for high-throughput community assays and molecular diagnostics, allowing for accurate identification of fungal taxa in various environments.

Certain protein-coding genes, such as those involved in metabolic pathways or stress responses, can also serve as conserved markers for molecular identification and phylogenetic studies of fungal species. These genes often exhibit conserved sequences, which can be used for amplification and sequencing, aiding in the identification of various beneficial fungi. Several specific genes have been widely utilized in fungal systematics, each offering unique advantages for distinguishing between species. One of the most commonly used protein-coding genes is Translation Elongation Factor 1-alpha (*TEF1-α*). This gene has been recognized for its efficacy in determining relationships among closely related fungal species due to its moderate level of variability, making it suitable for both deep and shallow phylogenetic analyses. Another important protein-coding gene is RNA Polymerase II Subunit B (*RPB2*). The *RPB2* is often used together with other genomic regions to enhance the resolution of phylogenetic trees. The *β-tubulin* gene is also frequently used as a molecular marker for fungal identification. This gene is particularly valuable for identifying species within the Ascomycota and Basidiomycota phyla, as it provides sufficient variation to differentiate between species and has a conserved structure. Additionally, the Calmodulin (*CaM*) gene has emerged as a secondary barcode or conserved genomic region in fungal identification. Its use can be helpful in cases where other markers may not provide clear distinctions, as it has been shown to be effective in resolving species boundaries in certain fungal taxa. Also, the glyceraldehyde-3-phosphate dehydrogenase (*gpd*) gene is a key molecular marker used for the identification and classification of fungi. This gene encodes an enzyme involved in the glycolytic pathway, which is essential for cellular metabolism in fungi. Its importance in fungal identification arises from its conserved structure across various fungal species, making it a reliable target for molecular assays.

The ITS region, although is not a protein-coding gene, is often mentioned alongside these markers due to its widespread use in fungal identification. It serves as a primary barcode for many fungi, complementing the information obtained from protein-coding genes. Integration of various protein-coding genes, such *as TEF1-α, RPB2, β-tubulin, CaM,* and *gpd*, together with the ITS region via multi-locus gene sequencing enhances the accuracy and reliability of fungal identification in Iran and worldwide, facilitating better understanding of diversity and phylogeny in fungal populations (Daroodi and Taheri 2024; Alijani Mamaghani et al. 2024). Classification of the soil-inhabitant fungi in different taxonomic groups revealed high level of diversity among these fungal communities and their various ecological roles in soil ecosystems. Diversity of soil fungi in Iran is influenced by the unique climatic and different geographical conditions in various regions, which necessitates localized studies to identify the most beneficial soil-inhabiting fungal species for specific agricultural situations.

Accurate identification and characterization of soil-inhabiting fungi is essential for their effective application in agricultural systems. Recent advancements in molecular techniques, such as DNA barcoding and metagenomics, have improved the accuracy of fungal identification. These methods allow for differentiation of various taxa and understanding their ecological functions in soil. For instance, a web-based platform, SoilFun, has been developed to facilitate the identification and classification of soil fungi, providing researchers with valuable resources for studying these different aspects related to the soil fungi.

In Iran, studies have been focused on isolating and identifying various types of soil-inhabiting fungi from different soil types and plant roots in various climates (Daroodi et al. 2022; Daroodi and Taheri 2023; Masoudi Khorasani et al. 2023). Investigations revealed that numerous species of fungi can enhance soil health, improve nutrient cycling, promote plant growth, and decrease the destructive effects of both biotic and abiotic stresses on various plants. In addition, some beneficial fungal species resident in soil are effective in producing bioactive metabolites themselves, or increasing accumulation of bioactive compounds in the host plant tissues, with high value and applications in medicine, agriculture and industry. Different types of soil-inhabiting beneficial fungi, including endophytes, mycorrhizae, and rhizospheric fungi obtained from various climatic regions of Iran are reviewed and discussed below.

3 Endophytic Soil-Inhabiting Fungi in Iran, Their Beneficial Effects and Modes of Action

Numerous researchers reported identification and beneficial effects of endophytic soil-inhabiting fungi obtained from below ground tissues of several plant species in Iran. According to the previous investigations, tropical and subtropical regions have higher abundance and diversity of endophytic fungi than in temperate and cold regions (Chen et al. 2023; Rana et al. 2019), as observed in the tropical regions of this country (Alijani Mamaghani et al. 2024). Endophytic fungi have numerous modes of action with various benefits. The endophytic fungus *Acrophialophora jodhpurensis* obtained from tomato roots in Iran is resident in soil with high biocontrol effect against *Rhizoctonia solani* AG4-HG II (Daroodi et al. 2021), *Alternaria alternata* (Daroodi et al. 2022), and *Fusarium graminearum* (Daroodi and Taheri 2023) in both *in vitro* and *in vivo* conditions. The mycelia of this endophytic fungus were extracted by methanol and the obtained extract was identified using different chromatography methods. Phytochemical analysis via thin-layer chromatography (TLC) and nuclear magnetic resonance (NMR) spectroscopy revealed that ergosterol peroxide was the main metabolite in the extract obtained from the mycelia of *A. jodhpurensis*. Antifungal activities of the methanolic extract and ergosterol peroxide in the culture media were investigated on *R. solani* AG4-HG II. Ergosterol peroxide showed antifungal effects on the hyphae of *R. solani* and induced formation of sclerotia by this phytopathogen (Daroodi et al. 2024). Histochemical analyses revealed that treatment with ergosterol peroxide induced apoptosis in the mycelia of *R. solani* via superoxide anion (O_2^-) accumulation and hyphal cell death. Application of ergosterol peroxide via treating tomato leaf discs reduced progress of the disease caused by *R. solani* on the leaves. Therefore, antifungal metabolites produced by the soil-inhabiting endophytic fungus *A. jodhpurensis*, such as ergosterol peroxide, are effective in controlling destructive phytopathogens, like *R. solani*, in both *in vitro* and *in vivo* conditions (Daroodi and Taheri 2024).

Endophytic fungi, such as *Penicillium canescens* and *Talaromyces* sp., obtained from the roots of *Salvia abrotanoides* (as a medicinal plant), considerably increased growth of the host plant and accumulation of active compounds, including cryptotanshinone and tanshinone IIA by 77.00% and 19.64%, respectively (Masoudi Khorasani et al. 2023). Cryptotanshinone is a quinone diterpene extracted from the roots of *Salvia* spp., which has considerable anticancer effects. Cryptotanshinone can reduce the tumor mass, increase the tumor suppression rate, and enhance the chemotherapeutic effect of 5-fluorouracil by a mechanism related to inhibition of the Janus kinase 2/signal transducer and activator of transcription 3 signaling pathway (Wu et al. 2022). Tanshinone is a class of abietane diterpene compounds isolated from *Salvia* spp. Tanshinone, also known as total tanshinone, is a lipid-soluble phenanthrenequinone compound, with antibacterial, anti-inflammatory, anticancer properties and other therapeutic effects for human diseases via a network of complex mechanisms (Wang et al. 2015). Symbiosis of various endophytic fungi with the root system of *Salvia* spp. can significantly increase accumulation of these medically important compounds extracted from the host plant (Masoudi Khorasani et al. 2023).

In a recent study, diversity of endophytic fungi obtained from three species of *Ferula* spp. (in the family Apiaceae) have been investigated in Iran. A total of 1348 endophytic strains were isolated from *Ferula ovina*, *F. galbaniflua,* and *F. persica.* The fungal strains belonged to Eurotiales (16 species), Pleosporales (11 species), Botryosphaeriales (1 species), Cladosporiales (2 species), Helotiales (6 species), Hypocreales (31 species), Sordariales (7 species), Glomerellales (2 species), and Polyporales (1 species). Therefore, the most of fungal endophytes obtained were within the phylum Ascomycota. *Ferula ovina* had the richest species composition of endophytic fungi, and the endophytic fungi were most abundant in the plant roots compared to the shoot system. The endophytic fungal communities obtained from each of *F. ovina* roots or shoots, *F. galbaniflua* roots, and *F. persica* roots were diverse species groups originating from a similar habitat. Their distribution followed a normal distribution, suggesting that the interactions of numerous independent environmental factors can be involved in the fungal species abundance. Principal component analyses revealed that the highest species diversity and dominance were observed in the endophytic fungal communities isolated from *F. ovina* and *F. persica* roots, respectively. However, beneficial effects of these endophytic fungi on plant growth and protection against environmental stresses have not been investigated yet, which can be interesting subjects for future research.

In a recent study performed by Alijani Mamaghani et al. (2024), diversity of *Alternaria* and *Fusarium* species, was investigated as the most abundant endophytic fungi isolated from healthy potato plants in two climatically different regions (Ardebil in the north-west and Kerman in the south-east) of Iran. Seventy-five endophytic *Fusarium* strains and 83 *Alternaria* strains were identified using multi-locus sequence analysis. *Alternaria* strains were characterized by the sequences of *gpd* and *CaM* genomic regions and the phylogenetic tree showed three well-separated clades. Among 83 *Alternaria* strains, 73 were placed in clade A which referred as Alternaria section, 6 strains were included in clade B which was Ulocladioides section, and 4 strains were in clade C referred as Infectoriae section. It seems that different climatic conditions in the two separate geographic regions of Iran (Ardebil and Kerman) do not affect significantly the distribution of endophytic *Alternaria*; in both regions the *Alternaria* section was predominant, apparently due to greater tolerance to diverse climate conditions.

Endophytic *Fusarium* spp. strains were identified via sequencing the translation elongation factor 1α (*tef1*), β-tubulin (*tub2*) and internal transcribed spacer (ITS) genomic regions. Molecular analyses revealed the presence of 13 *Fusarium* species, including *F. brachygibosum, F. clavum, F. equiseti, F. flocciferum, F. incarnatum, F. nirenbergiae, F. nygamai, F. oxysporum, F. proliferatum, F. redolens, F. sambucinum, F. solani* and *F. thapsinum.* In particular, *F. nygamai* was predominant in south Kerman, and *F. equiseti, F. oxysporum* and *F. solani* were more frequently obtained from Ardebil (Alijani Mamaghani et al. 2024). So, the strains of *F. nygamai* were more adapted to dry and hot climate conditions in Kerman province of Iran, whereas *F. oxysporum* and *F. solani* can apparently colonize potato plants under diverse climatic conditions.

Among the endophytic *Fusarium* spp. strains, 26 strains of *F. equiseti, F. nirenbergiae, F. oxysporum, F. nygamai, F. proliferatum*, and *F. sambucinum*, were selected for investigating their ability in production of various mycotoxins such as deoxynivalenol (DON), nivalenol (NIV), diacetoxyscirpenol (DAS), T-2 toxin (T-2), beauvericin (BEA), enniatins (ENNs), fumonisins (FBs), fusaric acid (FA) and moniliformin (MON). None of the tested endophytes produced trichothecenes (DON, NIV, DAS and T-2). Two out of 2 *F. equiseti* isolates, 1/6 *F. oxysporum*, 1/3 *F. proliferatum*, and 1/9 *F. nygamai* strains did not produce any of the toxins tested. The rest of endophytic *Fusarium* spp. strains produced one or more BEA, ENNs, FBs, FA and MON toxins. The most toxigenic strain, *F. nygamai* ITEM-19012, produced the highest quantities of FBs (7946, 4693 and 4333 μg/g of B1, B2, and B3 respectively), along with the highest quantities of both BEA and MON (Alijani Mamaghani et al. 2024). These findings suggest that contamination of potato tubers (colonized by endophytic fungi without any disease symptoms) with mycotoxins in the field or postharvest due to a change in lifestyle of endophytes to pathogens should be carefully considered. There are previous reports demonstrating that under specific conditions especially when the plant is under stress, endophytic fungi colonizing in the healthy plant tissues may switch from non-pathogenic to pathogenic manner (Hardoim et al. 2015).

It is not well-known that the endophytic *Fusarium* spp. may switch to a pathogenic lifestyle and, in this case, if accumulation of toxins in the edible tubers can occur in the field or during storage period. Classical biological control is based on the concept that most effective biocontrol agents should be sought amongst the coevolved antagonists of the target pathogen, which are supposed to have the highest specificity and biocontrol efficacy. Determining effects of these endophytic fungi (obtained from potato in Iran) on plant growth factors, biotic and abiotic stresses management and exploring their modes of action can be interesting subjects for future research.

In another large-scale investigation on endophytic fungi of potato plants in Iran performed by Alijani Mamaghani et al. (2022), of 922 endophytic fungal isolates obtained from different parts of potato plants including leaf, stem, root and tuber, 33 of them (3.57%) belonged to the genus *Cephalotrichum*. Of these endophytes, eight strains were obtained from Ardebil which is a temperate and cold region and 25 from the south Kerman province as a tropical zone; most of them were isolated from the tubers. Morphological and molecular investigations (using ITS, *tub2*, *tef1* and LSU genomic regions analyses) revealed the presence of endophytic fungi, such as *C. asperulum*, *C. gorgonifer* and *C. tenuissimum*, in potato plants. These researchers also identified the fungal isolates using matrix-assisted laser desorption/ionization time of flight (MALDI-TOF) mass spectrometry (MS) and the spectra obtained were used to construct Superspectra for MALDI-TOF MS identification of these endophytic species. They reported that MALDI-TOF MS is a reliable, timesaving and cost-effective method, which can be used for identification of different fungal taxa (Alijani Mamaghani et al. 2022).

The species *C. gorgonifer* was the most frequently isolated among all species obtained in their study and it was recovered from both sampling areas, whereas *C. gorgonifer* was reported only from Ardebil province. The strains of *C. asperulum* and *C. tenuissimum* originated from potato plants in south Kerman province, therefore they are apparently most common in dry and hot climates (Alijani Mamaghani et al. 2022). Endophytic *Cephalotrichum* spp. may be dormant and wait for their host to senesce to start decomposition of its tissues, to finally become active saprophytes as soon as the tissues they have colonized as endophytes become senescent. As lignin is a major structural component often limiting decomposition, these ligninolytic endophytes are of particular importance in terms of their role in carbon turnover and nutrient cycling in ecosystems (Osono and Hirose 2011) and further studies seems to be necessary on their ecology and function in sustainable agriculture.

Endophytic fungal isolates were obtained from leaves, pods, stems and roots of aerobic rice plants in Golestan province of Iran. These isolates were identified via morphological and molecular methods and belonged to several species, including *Fusarium chlamydosporum, F. proliferatum, F. incarnatum, F. solani, F. verticillioides, Nigrospora sphaerica, N. oryzae, Epicoccum nigrum,* and *Cladosporium cladosporioides.* Biocontrol effects of these endophytic fungi were investigating against *Pyricularia oryzae* and *Fusarium fujikuroi,* the causal agents of rice blast and crown and root rot diseases, respectively. In the *in vitro* assays, three methods, including dual culture, volatile metabolite and non–volatile metabolite bioassays were used to determine the most effective endophytes against each pathogen. Their findings revealed that, in dual culture method, the *F. incarnatum* had the highest inhibition rate against the growth of *P. oryzae* and *F. fujikuroi*, with 58.64 and 60.03% inhibition rates, respectively. In volatile metabolites method, *E. nigrum* with 57.13% and *F. incarnatum* with 52.24% inhibition showed the highest percentages of inhibition for *F. fujikuroi*. In the non–volatile metabolites assay, *F. incarnatum* and *F. proliferatum* showed the highest percentages of inhibiting the growth of *P. oryzae*. The endophytic *F. incarnatum* showed antagonistic activity in all three *in vitro* experiments, including dual culture, volatile and non–volatile metabolites bioassays against both phytopathogenic fungi tested. In the *in vivo* investigations, the isolates of *F. proliferatum* and *F. incarnatum* showed the best biocontrol effect against both diseases and increased the height and fresh weight of root system and aerial parts of rice plants under greenhouse conditions. However, detailed modes of action of these beneficial fungal endophytes in plant defense induction against the pathogens and growth-promotion should be investigated in future research.

Trichoderma species are another group of soil inhabiting beneficial fungi, which are frequently isolated and reported from the soil and plant organs in different climatic regions of Iran and have worldwide distribution (Bazgir et al. 2023; Ghasemi et al. 2019; Hajieghrari et al. 2008). The genus *Trichoderma* consists of globally distributed fungi. Among them, *T. harzianum*, is one of the most commonly collected *Trichoderma* species, which had been known as an aggregate species. In a recent study, Bazgir et al. (2023) reported identification of *Trichoderma* spp. strains belonging to 5 species of the Harzianum clad, which were isolated from soil samples collected from diverse geographical regions of Khorramabad region in Iran. Comparing the cultural and phenotypic criteria combined with molecular analysis of the *tef1*-α gene sequences of the isolated fungi, the species including *T. afroharzianum*, *T. atrobrunneum*, *T. guizhouense*, *T. harzianum* and *T. pholiotae* were identified in their study. The strains of *T. harzianum* had the highest frequency among the 5 species reported by Bazgir et al. (2023). However, beneficial effects of these *Trichoderma* spp. strains on plant growth-promotion and resistance induction against various biotic and abiotic stresses were not investigated yet, which can be interesting subject for future research.

Various *Trichoderma* species have been reported to be effective in suppressing stem, crown and root diseases caused by soil-borne phytopathogens, mainly including fungi, Oomycetes and nematodes (Intana et al. 2023; Legrifi et al. 2024). For instance, root rot disease caused by *Pythium schmitthenneri* is an important challenge in olive (*Olea europaea* L.) cultivation, which can be biologically controlled via application of *Trichoderma* spp. as a sustainable alternative to chemical fungicides (Legrifi et al. 2024). Four species, including *T. harzianum*, *T. longibrachiatum*, *T. virens*, and *T. viride*. were obtained from olive roots, among them, *T. harzianum* T-BM6 and *T. virens* T-KH4 showed the highest inhibitory effects against *P. schmitthenneri*. Genomic analysis highlighted key genes related to the MAPK signaling pathway, sulfur metabolism, inositol phosphate metabolism, phenylalanine, tyrosine, and tryptophan metabolism, which are involved in beneficial effects of these fungal species (Legrifi et al. 2024).

The findings of numerous researchers revealed that several species of *Trichoderma*, such as *T. harzianum, T. virens, T. asperellum,* etc. are promising candidates for biological control (Intana et al. 2023; Legrifi et al. 2024; Samlikamnoed et al. 2023) offering a sustainable alternative to chemical treatments for managing plant disease. More detailed focus on the role of various *Trichoderma* species in controlling phytopathogens *in vitro* and *in vivo* conditions, their potential in plant protection against abiotic stresses, and also formulation and application strategies to increase stability and efficiency of these biocontrol agents seems to be necessary in future research.

Numerous endophytic fungi have been obtained and identified from medicinal plants, such as *Salvia multicaulis,* in different regions of Iran.

Studies of Sabet Jahromi et al. (2021) on the fungal endophytes isolated from *S. multicaulis* plants in their main wild habitats in Iran, including Qazvin, Alborz and Mazandaran provinces, revealed the presences of endophytic fungi belonging to numerous taxa. Their findings indicated that Ascomycota was the dominant taxon in colonizing this medicinal plant species with relative frequency of 96.77%, comprising of Eurotiomycetes (40.5%), Sordariomycetes (33.9%) and Dothideomycetes (20.5%). Mucoromycota, comprised the rest of endophytes (5.23%). Their investigation revealed the presence of several fungal genera including *Fusarium*, *Penicillium*, *Aspergillus*, *Alternaria*, *Colletotrichum*, *Rhizopus*, *Macrophomina*, *Trichoderma* and *Nodulisporium* in the endophytic microbiome studied (Sabet Jahromi et al. 2021) as listed in Table 8.1. They reported the highest endophytic fungi isolation in summer, in warmer regions and from the roots (91%), compared to the aerial parts of plants investigated for the presence of endophytic fungi. Several other reports about fungal endophytes from Iran in recent years are listed in this table.

4 Rhizospheric and Mycorrhizal Fungi of Iran and Their Mechanisms of Action in Sustainable Agriculture

Rhizosphere is the narrow region of soil surrounding plant roots and is directly influenced the root system. This area is characterized by the interactions between the roots, soil, and various microorganisms, mainly fungi. In the rhizosphere, root exudates play a crucial role in shaping the microbial communities and nutrient dynamics. These secretions can alter the chemical and biological properties of the surrounding soil, promoting the growth of beneficial microorganisms while inhibiting harmful phytopathogens. This interaction is vital for nutrient uptake, as the rhizosphere is where plants access essential minerals and water. The rhizosphere is not just a passive environment; it is a dynamic zone where complex biochemical processes occur. The presence of roots affects soil structure, moisture retention, and the

Table 8.1 Some of the endophytic fungi obtained from plant roots reported from Iran in recent years and their beneficial effects on plants

Endophytic fungi	Host plant	Region	Beneficial effects on plant growth and against stresses	Mode of action	References
Fusarium, Penicillium, Aspergillus, Alternaria, Colletotrichum, Rhizopus, Macrophomina, Trichoderma and Nodulisporium	*Salvia multicaulis*	Qazvin, Alborz and Mazandaran provinces	ND	ND	Sabet Jahromi et al. (2021)
Acrophialophora jodhpurensis	*Solanum lycopersicum*	Mashhad	Biocontrol of *Rhizoctonia solani, Alternaria alternata* and *Fusarium graminearum*	Extracellular enzymes, siderophore production, Phosphate solubilization, non-volatile metabolites, indole-acetic acid production, plant defense priming	Daroodi and Taheri (2023, 2024), Daroodi et al. (2021), Daroodi et al. (2022)
Paecilomyces maximus	*Capparis spinosa*	Zanjan	ND	ND	Rajabi et al. (2022)
Fusarium chlamydosporum *F. proliferatum* *F. solani* *F. fujikuroi*	*Oryza sativa*	Golestan	ND	ND	Zare et al. (2023)
Fusarium sp. *Trichoderma* sp. *Alternaria* sp.	*Euphorbia esula* *Chenopodium album*	Hamedan	ND	Extracellular enzymes, secondary metabolites	Bakhshiani et al. (2020)
Aspergillus niger , *Paecilomyces formosus,*	*Pistacia mutica*	Hormozgan	Salinity stress tolerance	ND	Kavehnia et al. (2023)
Fusarium, Colletotrichum, Phomopsis	*Taxus baccata*	Gorgan	Biocontrol of Fusarium oxysporum	ND	Ashnavar et al. (2022)

Beauveria bassiana	*Brassica napus*	Tehran	ND	ND	Seyed Talebi et al. (2018), Solhjouy-Fard et al. (2023)
Alternaria alternata Cladosporium herbarum Paraphaeosphaeria sporulosa, Neocucurbitaria unguis Pleosporales sp. *Fusarium equiseti* *Fusarium fujikuroi* *Gnomoniopsis idaeicola Epicoccum nigrum*	*Corylus avellana*	Six provinces of Iran	ND	ND	Kashanian et al. (2021)

ND not determined

availability of nutrients, which in turn influences plant health and productivity. The microorganisms in this region, including numerous fungal species, contribute to nutrient cycling and organic matter decomposition, resulted in enhancing soil fertility. So, the rhizosphere is a critical component of different ecosystems, serving as a bridge between plants and their soil environment, with an important role in agricultural productivity and sustainability.

Rhizospheric fungi play a crucial role in enhancing agricultural yield in various climatic regions, worldwide. In arid and semi-arid regions, such as those found in central and eastern Iran, specific fungal species have been identified that help plants cope with water stress and improve nutrient uptake, particularly phosphorus and nitrogen. For instance, mycorrhizal fungi are known to form symbiotic relationships with many crops, enhancing their resilience to drought and salinity, which are prevalent challenges in these areas. Additionally in temperate regions, rhizospheric fungi contribute to not only protect plants against abiotic stresses, but also are involved in biological control of plant pathogens which leads to reducing the reliance on chemical pesticides and promoting sustainable agricultural practices. Diversity and functional roles of these fungi underscore their importance in maintaining soil health and supporting crop yields in Iran's wide range of climatic conditions, making them vital for sustainable agricultural development.

Several researchers reported isolation and identification of fungi obtained from the rhizosphere of different plant species in various regions of Iran. For instance, Alidadi et al. (2020) demonstrated morphological and molecular identification of several rhizospheric fungal species via sequencing of rDNA-ITS region. The fungi obtained in their research, including *Mortierella alpina, Penicillago nodositata*, and *Trichocladium jilongensis,* were isolated from oak trees rhizosphere in Zagros forests, located at Ilam province (Table 8.2). However, effects of these fungal strains on growth factors and stress protection of the host plants have not been studied yet, which can be investigated in future research.

Arbuscular mycorrhizal fungi (AMF), which can be obtained from the rhizosphere, play a crucial role in enhancing soil health and plant growth. These fungi form symbiotic relationships with plant roots, facilitating nutrient uptake, especially phosphorus, and improving soil structure. Several studies reported that AMF can significantly influence the microbial community dynamics in the rhizosphere, which is essential for maintaining soil fertility and plant health under various environmental conditions, including arid and semi-arid regions typical of Iran. Furthermore, studies revealed that the presence of AMF can induce defense responses of plants to both biotic and abiotic stresses. This symbiotic relationship is particularly beneficial in degraded lands, where AMF can aid in the restoration of soil quality and support sustainable agricultural practices.

Numerous studies demonstrated the presence of several taxa of AMF species in various regions of Iran. Studies related to the AMF strains in this country are mainly focused on morphological and molecular identification, also on demonstrating diversity and abundance of this important group of soil inhabiting beneficial fungi. Some studies reported their modes of action in the symbiotic interaction with the host plants and their effect in plant protection or growth-promotion. For example,

Table 8.2 Some of the rhizospheric and arbuscular mycorrhizal fungi reported from Iran and their beneficial roles in agriculture

Rhizospheric and arbuscular mycorrhizal fungi	Host plant	Region	Beneficial effects on plant growth and against stresses	Mode of action	References
Rhizophagus intraradices *Funneliformis mosseae*	Chive (*Allium fistulosumo)*	Rafsanjan (Kerman province)	Increased plant growth index	Increased absorption of phosphorus and nitrogen/ increased chlorophyll, copper, iron, manganese and zinc elements	Ilkhan et al. (2023)
Acaulospora, Archaeospora, Entrophospora, Gigaspora, Claroideumglomus,Funeliformis, Rhizoglomus, Septoglomus and *Scutellospora*	Wheat root and rhizospheric soil	Kohgiluyeh and Boyer-Ahmad province (in the southwestern region of Iran)	Increased plant growth and phosphorous uptake	Increasing solubility of phosphorous and improving the soil texture and structure	
Mortierella alpina *Penicillago nodositata* *Trichocladium jilongensis*	Rhizosphere of oak trees	Zagros forests (Ilam province)	ND	ND	Alidadi et al. (2020)
Funneliformis mosseae *Acaulospora spinosa*	Alfalfa	Kohgiluyeh & Boierahmad province	Reduced severity of take-all disease caused by *Gaeumannomyces graminis* var. *tritici* and improved wheat growth indices	ND	Sadravi (2012), Nikmaram et al. (2024)
Funneliformis mosseae	Pistachio	–	Increase the humidity of leaves and root volume, plant protection against salinity and drought stress	ND	Fattahi et al. (2021)
Funneliformis mossea, Rhizoglomus intraradies *Blaszkowskia deserticola*	Wheat	–	Reducing the severity of the disease, biocontrol of the fungal phytopathogen *Gaeumannomyces graminis* var. *tritici*	Induction of defensive reactions	Nikmaram et al. (2024)

(continued)

Table 8.2 (continued)

Rhizospheric and arbuscular mycorrhizal fungi	Host plant	Region	Beneficial effects on plant growth and against stresses	Mode of action	References
Glomus mosseae	*Mellisa officinalis*	–	Reduced symptoms of yellowing and wilting caused by Fusarium culmorum, increased growth index	Increased activity of antioxidant enzymes (SOD, CAT, APX, POX, PPO) and increased accumulation of non-enzymatic antioxidants (carotenoid, ascorbate, and phenolics), increase PAL and LOX activities	Saderi et al. (2024)
Rhizophagies intraradies	Saffron	–	Increase dry yield of stigma (0.0242 g pot $^{-1}$)	Increase absorption of N, K and P	Habibi et al. (2021)
Claroideoglomus etunicatum	Corn	–	Maintain plant growth, plant protection against drought stress	Peroxidase polyphenol oxidase PAL	Sedaghati et al. (2021)
G. intraradices *G. fulvum* *G. constrictum* *G. sinuosum* *G. versiforme*	Wheat	Central regions and northwest of Iran	Plant protection against salinity	ND	
Glomus dimorficum, *Glomus etunicatum*, *Glomus macrocarpum* *Glomus geosporum* *Glomus fasciculatum*	Barley and wheat	Central regions and northwest of Iran	Plant protection against drought	ND	Jahromi et al. (2012)

Glomus aggregatum *Glomus mosseae* *Glomus intraradices* *Glomus geosporum* *Glomus fasciculatum* *Glomus manihotis* *Glomus macrocarpum* *Glomus albidum* *Paraglomus occoltum* *Gigaspora albida*	Rhizospheric soil of several plant species	Kiasar forest, Damghan	ND	ND	Modarresi Chahardehi et al. (2014)
Glomus nanolumen	Rhizosphere of mulberry trees (*Morus alba*)	Tehran	ND	ND	Zangeneh (2022)
Acaulospora alpina *A. gedanensis* *A. herrerae* *A. scrobiculata*	Rhizosphere of feverfew (*Tanacetum parthenium*) and wheat	Isfahan, Sistan and Baluchestan provinces	ND	ND	Zanganeh (2021)
Rhizophagus aggregatus *Funneliformis geosporus* *Rhizophagus fasciculatus* *Funneliformis caledonius* *Funneliformis mosseae* *Claroideoglomus etunicatum* *Glomus macrocarpum* *Glomus versiforme* *Septoglomus constrictum* *Glomus glomerulatum* *Scutellospora calospora*	Rhizospheric soil of grape producing areas	West Azerbaijan province (in the northwest of Iran)	Effective on soil phosphorus contents	The AMF spore density, AMF colonization frequency and intensity were negatively correlated with soil phosphorus contents	Rezaee Danesh et al. (2022)
Funneliformis coronatum *Funneliformis mosseae* *Septoglomus constrictum* *Gigaspora gigantea* *Scutellospora calospora*	Rhizosphere of onion, alfalfa, corn, almond, peach, apricot, and sunflower	Rafsanjan	ND	ND	Aminiannasab et al. (2021)

(continued)

Table 8.2 (continued)

Rhizospheric and arbuscular mycorrhizal fungi	Host plant	Region	Beneficial effects on plant growth and against stresses	Mode of action	References
Glomus mosseae *Hebeloma sinapizans*	Oak roots	Forests in the northern regions of Iran	Increased stem biomass, leaf biomass, leaf area in the seedlings inoculated with *H. sinapizans*	Increased chlorophyll and water content in the host tissues	Zamani and Matinizadeh (2021)
Glomus macrocarpum *G. caledonium* *G. clarum* *G. constrictum* *G. etunicatum* *G. deserticola*	Rhizospheric soil and roots of medicinal plants	Tandore national park, north of Iran	Increased plant biomass	Absorption of more nutrients, including phosphorus and potassium, in the plant	Esmaeilzadeh et al. (2006a, b)
Glomus aggregatum *G. etunicatum* *G. fasciculatum* *G. occultum* *G. mosseae* *G. intraradices* *G. geosporum* *G. diaphanum* *Scutellospora callospora*	Rhizosphere of thyme	Tehran	ND	ND	Ahmadi et al. (2016)
Rhizophagus iranicus *Claroideoglomus claroideum* *Funneliformis mosseae*	Pistachio roots	Kerman province	ND	ND	Yazdan Panah et al. (2017)
Glomus mosseae	Wheat rhizosphere	Damghan	ND	ND	Eskandari and Danesh (2012)
Funneliformis mosseae, Funneliformis caledonius, Rhizophagus intraradices, Claroideoglomus drummondii, Gigaspora gigantea	Roots and rhizosphere of *Amygdalus scoparia* (almond)	Zagros forests in Ilam province	ND	ND	Mirzaei and Noorbakhsh (2015)

ND not determined

Rezaee Danesh (2013) demonstrated identification of the AMF species obtained from the rhizosphere of barley plants in Damghan region (with tropical climate), together with determining spore population abundance and their diversity pattern. The spore population was 72–840 spores per 35 g dry soil. Different AMF species were identified in the barley rhizosphere, which 13 species belonged to *Glomus* spp. and only 1 species was identified in each of the *Acaulospora, Pacispora* and *Scutellospora* genera. Therefore, *Glomus* was the prevalent genus in barley rhizosphere in Damghan region. The strains of *G. corymbiforme* and *G. trimurales* were reported for the first time in Iran. The most and least relative abundances were observed for the strains of *G. intraradices* (12.66%) and *Pacispora scintillans* (1.99%), respectively (Rezaee Danesh 2013). Future research can be focused on the modes of action of these AMF species in interaction with various plant species and their efficacy in agricultural systems to increase plant production and enhance crop immunity against environmental stresses.

In a study on AMF flora of alfalfa rhizosphere in Kohgiluyeh and Boierahmad province located in southwest of Iran, 23 fungal species were identified belonging to nine genera, including *Acaulospora, Claroideoglomus, Diversispora, Entrophospora, Funneliformis, Gigaspora, Glomus, Septoglomus* and *Scutellospora*. Among them, *F. mosseae* was the most abundant species (Sadravi 2012). Biocontrol effect of some AMF strains reported by Sadravi (2012) was recently demonstrated against take-all disease on wheat plants caused by *Gaeumannomyces graminis* var. *tritici* (Nikmaram et al. 2024). Investigating effect of three AMF strains, including *Funneliformis mosseae*, *Rhizoglomus intraradices*, and *Blaszkowskia deserticola* alone and in combination on severity of the take-all disease and growth indices of wheat plants under greenhouse conditions showed that all AMF strains tested decreased the disease progress and increased growth factors compared to the control plants. The strain of *F. mosseae* was more effective in plant growth-promotion than other AMF strains used in the research performed by Nikmaram et al. (2024).

Ilkhan et al. (2023) demonstrated the effect of *Rhizophagus intraradices* and *Funneliformis mosseae* on some growth and physiological characters and the uptake of some nutrients by scallion seedlings under greenhouse conditions. The investigated factors included the mixed application of arbuscular mycorrhizal species in three levels (seed treatment, soil treatment and control) and soil sterilization at two levels (sterile and non-sterile). The obtained results revealed that all the investigated growth factors significantly increased in the scallion seedlings inoculated with the AMF strains compared to non-mycorrhizal plants (Table 8.2). The highest fresh and dry weights of shoots and roots were obtained by mycorrhizal seed treatment and soil treatment in sterile soil, respectively. The highest shoot height, phosphorus concentration and nitrogen uptake of scallion seedlings were observed in the seedlings treated by the AMF strains via soil treatment in sterile soil. The mycorrhizal colonization, the amount of copper, zinc, manganese and iron concentrations, chlorophyll and carotenoid contents increased significantly in AMF treated scallion seedlings via seed treatment compared to other types of treatments tested (Ilkhan et al. 2023). Therefore, application of the AMF strains via seed coating can be a suitable method to improve the plant growth via increasing various nutrients absorption and biosynthesis of photosynthetic pigments.

Mycorrhizal symbiosis can activate plant defense responses against many phytopathogens, such as fungi. Recent research conducted by Saderi et al. (2024) demonstrated the effect of pre-inoculation of lemon balm with the AM fungus *Glomus mosseae* on progress of the disease caused by *Fusarium culmorum*. The findings revealed that mycorrhizal colonization significantly reduced the disease index on the leaves and roots by 46% compared to non-mycorrhizal plants. Priming the plant defense response was observed in lemon balm pre-inoculated with AMF compared to the non-mycorrhizal control plants. The activity of antioxidant enzymes, including superoxide dismutase (SOD), catalase (CAT), ascorbate peroxidase (APX), peroxidase (POX), and polyphenol oxidase (PPO), as well as the contents of non-enzymatic antioxidants (carotenoid, ascorbate, and phenolic compounds) were higher in the AMF pre-treated plants infected with the pathogen than in non-mycorrhizal controls. Stronger antioxidant accumulation and activities led to decrease in the pathogen-induced hydrogen peroxide (H_2O_2) and malondialdehyde (MDA) accumulation in mycorrhizal plants. This symbiotic interaction increased phenylalanine ammonia lyase (PAL) and lipoxygenase (LOX) activities, which resulted in the accumulation of salicylic acid (SA) and jasmonic acid (JA). Findings of Saderi et al. (2024) indicated that *G. mosseae* colonization of lemon balm activated the plant defense against *F. culmorum* infection via recovering photosynthesis rate, strengthening antioxidants machinery, and triggering SA and JA signal transduction pathways.

Also, several other mechanisms might be involved in positive effects of the AMF species and other beneficial soil-inhabiting fungi on plant growth-promotion and defense induction against both biotic and abiotic stresses as demonstrated in numerous reports (Ibiang et al. 2023; Sellappan et al. 2023; Roy et al. 2024). Mechanisms of action of beneficial fungi resident in soil (including endophytics, rhizospherics and mycorrhizae) on various physiological, biochemical, morphological, cellular and molecular processes in the host plants, leading to plant growth-promotion and defense activation against environmental stimuli are illustrated in Fig. 8.1.

In a recent study performed, the correlation between population and diversity of AMF and physicochemical characteristics of soil in wheat fields was investigated in Kohgiluyeh and Boyer-Ahmad province located in southwestern region of Iran. The AMF spores were isolated by sieving the rhizospheric soil suspension in water and centrifuging in sucrose solution. The spore population of these fungi was counted, and percentage of wheat root colonization was investigated microscopically for each sample. They found several AMF species belonging to nine genera, including *Acaulospora, Archaeospora, Entrophospora, Gigaspora, Claroideumglomus, Funeliformis, Rhizoglomus, Septoglomus* and *Scutellospora*. The AMF fungus *F. mosseae* was dominant species with 90% relative frequency in the regions investigated. The population of these fungi had a negative correlation with soil pH but showed positive correlation with soil soluble phosphorus and plant phosphorus content. Also, positive correlation was observed between the amount of sand in the soil and the root length colonization by AMF species. Therefore, the AMF species have

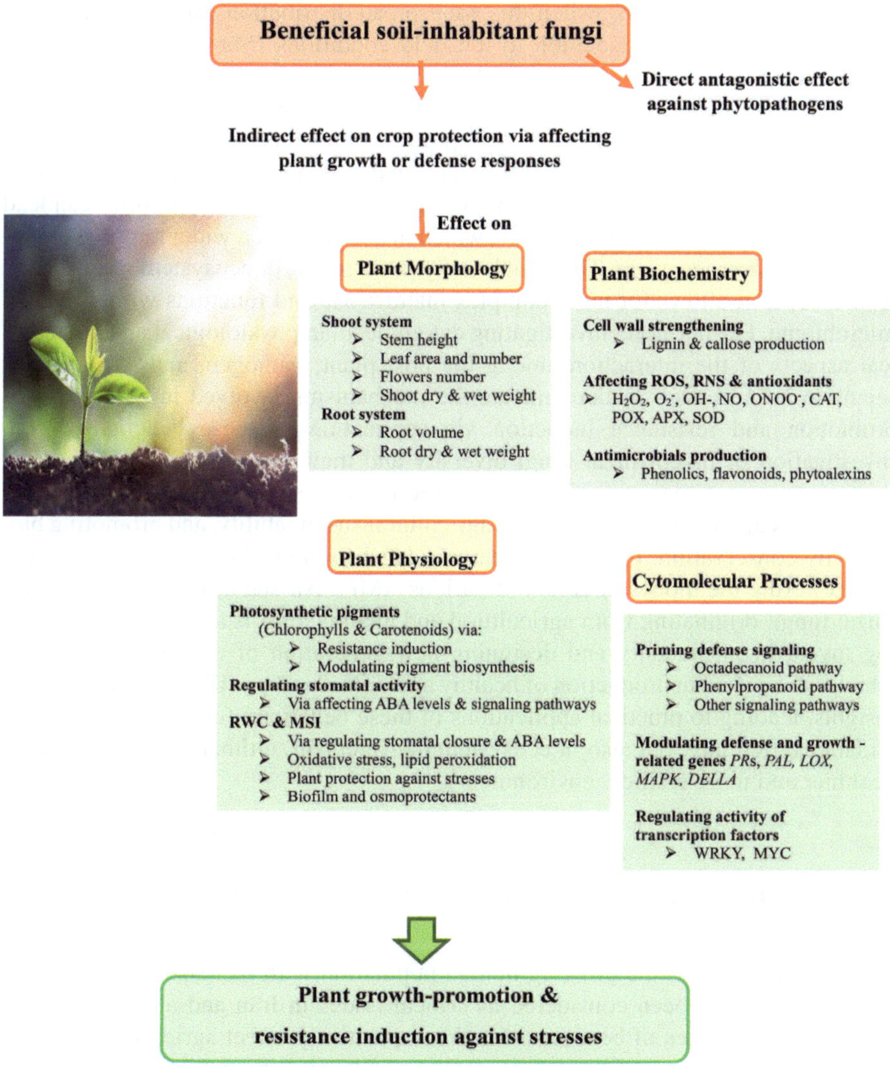

Fig. 8.1 Effect of beneficial soil-inhabitant fungi (endophytics, rhizospherics and mycorrhizae) on various physiological, biochemical, morphological, cellular and molecular processes in the host plants, leading to plant growth-promotion and defense activation against biotic and abiotic stresses

a higher population and diversity in soils with light texture and the positive correlation between population of the AMF species and plant phosphorus indicates their efficacy in uptake of phosphorous by the host plant.

Usually, agricultural systems have lower level of diversity within the AMF populations than natural environments and these systems are often dominated by a few taxa of the mycorrhizal fungi (Verbruggen et al. 2012). For instance, a specific genotype of *F. mosseae* was found to have a global distribution (in agricultural systems of Iran and other regions), which this widespread distribution can be related to its spread and successful adaptation to the field conditions (Marrassini et al. 2024; Nikmaram et al. 2024).

Diversity of mycorrhizal fungi is crucial for maintaining soil health and ecosystem sustainability. Therefore, more in depth research is needed to determine distribution and diversity of various AMF species and explore the factors affecting their diversity. Diverse AMF and other types of mycorrhizal fungi communities can lead to increased plant health and diversity, which in turn support a wider range of organisms within the ecosystem. This biodiversity is essential for ecosystem stability and resilience, as it allows for more complex interactions and functions within the soil microbiome. Furthermore, investigating cytomolecular, physiological and biochemical aspects of the interaction among the host plant, pathogen, and AMF strains seems to be necessary to demonstrate the mechanisms involved in plant growth-promotion and resistance induction via application of these beneficial fungi. Investigation of mycorrhizal fungi diversity and their mechanisms of interaction with plants in association with biotic or abiotic stresses is essential for advancing our ecological knowledge, improving agricultural sustainability, and promoting biodiversity conservation. Focus on research related to the isolation, characterization and exploring the modes of action of various AMF taxa and other types of mycorrhizal fungi, originating from agricultural and natural soils, is a key step in exploiting mycorrhizal diversity and designing new generation of microbial inoculants which can be used in production of healthy and organic crops. These studies provide insights, leading to practical applications of these beneficial fungi not only in sustainable agriculture, but also in ecosystem management, ultimately contributing to healthier and more resilient environments.

5 Concluding Remarks and Future Perspectives

Climate changes and the growing human population with increasing demands for food sources have been considered as critical issues in Iran and also on a global scale. Different types of beneficial fungi can positively affect agricultural systems via bioremediation, plant growth-promotion, and crop protection against various environmental stresses. Application of novel and effective formulations of these bioagents to maintain and improve their efficacy can increase agricultural products in different geographical regions with diverse climates.

Advanced Omics procedures, including genomics, transcriptomics, proteomics, and metabolomics, play key roles in enhancing application of beneficial fungi in agriculture. These methods can be used for comprehensive understanding of plant-microbe interactions, which is essential for developing effective bio-stimulants, bio-fertilizers and producing engineered beneficial microbes with higher levels of useful traits. Biotechnology and nanotechnology are increasingly used to engineer beneficial fungi with enhanced characteristics for organic plant production. Through genetic engineering, fungal genomes can be modified to improve their abilities to promote plant growth, enhance nutrient uptake, control pathogens and pests, and induce resistance against various environmental stimuli. This approach will be helpful in developing fungi which can thrive in various environmental conditions, thereby supporting crop resilience and productivity. Additionally, nanotechnology plays a crucial role in optimizing delivery of these fungi and their metabolites, ensuring that they effectively enhance the plant and soil health. By using nano-enabled systems, the targeted application of beneficial fungi can lead to improved soil fertility and reduced reliance on chemical fertilizers and pesticides, contributing to more sustainable agricultural practices. Application of these advanced technologies to increase efficacy of beneficial fungi can be interesting subject for future research, with considerable potential for enhancing agricultural sustainability and food security, worldwide.

References

Ahmadi T, Bernard F, Zangeneh S, Rejali F (2016) Identification of mycorrhizal fungi in the rhizospher of *Thymus daenensis* and mycorrhization of this species with *Glomus intraradices* in greenhouse conditions. J Plant Res (Iran J Biol) 28(4):674–683

Alidadi A, Kowsari M, Salehi Jozani G, Karimi E, Javan Nikkhah M (2020) New records of fungi from rhizosphere of oak trees for Iran fungal flora. Rostaniha 21(1):27–37

Alijani Mamaghani N, Saremi H, Javan-Nikkhah M, Respinis SD, Pianta E, Tonolla M (2022) Endophytic *Cephalotrichum* spp. from *Solanum tuberosum* (potato) in Iran—a polyphasic analysis. Sydowia 74:287–301

Alijani Mamaghani N, Masiello M, Somma S, Moretti A, Saremi H, Haidukowski M, Altomare C (2024) Endophytic *Alternaria* and *Fusarium* species associated to potato plants (*Solanum tuberosum* L.) in Iran and their capability to produce regulated and emerging mycotoxins. Heliyon 10(5):e26385

Aminiannasab P, Sedaghati E, Hosseini S, Saberi Riseh R (2021) Identification of arbuscular mycorrhizal fungi symbiotic with some plants in Rafsanjan city based on morphological features and sequence of beta-tubulin region. Microbial Biol 10(38):27–40

Ashnavar M, Rahnama K, Sarmast MK, Ghasemnezhad A (2022) The role of endophytic fungi isolated from native Iranian yew (*Taxus baccata*) in biological control of *Fusarium oxysporum*. Ecol Iran For 10(20):1–10

Badakhshan N, Mansouri SM, Habibi A, Naseri B, Shakeri S (2022) Effect of *Trichoderma harizianum* strain T22 on induced resistance of tomato to *Helicoverpa armigera* Hübner. J Iran Plant Prot Res 36(3):313–325. (In Persian with English abstract). https://doi.org/10.22067/JPP.2022.76074.1086

Bakhshiani M, Nazeri S, Zafari DM, Mahdian S (2020) Study on the presence of extracellular enzymes in culture of endophytic fungi isolated from the euphorbia esula L. and chenopodium album L. Agric Biotechnol J 12(3):141–156

Bazgir E, Zafari D, Darvishnia M (2023) Isolation and identification of Harzianum clade species of Trichoderma from Khorramabad County. Mycologia Iranica 10(2):67–78

Chen YJ, Chen HJ, Chung WH (2023) Endophytic fungal diversity in *Cirsium kawakamii* from Taiwan. J Fungi 9(11):1076

Daroodi Z, Taheri P (2023) Function of the endophytic fungus *Acrophialophora jodhpurensis*, methionine, and nitric oxide in wheat resistance induction against *Fusarium graminearum* via interplay of reactive oxygen species and iron. Physiol Mol Plant Pathol 128:102132. https://doi.org/10.1016/j.pmpp.2023.102132

Daroodi Z, Taheri P (2024) The genus *Acrophialophora:* history, phylogeny, morphology, beneficial effects and pathogenicity. Fungal Genet Biol 171:103875

Daroodi Z, Taheri P, Tarighi S (2021) Direct antagonistic activity and tomato resistance induction of the endophytic fungus *Acrophialophora jodhpurensis* against *Rhizoctonia solani*. Biol Control 160:104696. https://doi.org/10.1016/j.biocontrol.2021.104696

Daroodi Z, Taheri P, Tarighi S (2022) *Acrophialophora jodhpurensis*: an endophytic plant growth promoting fungus with biocontrol effect against *Alternaria alternata*. Front Plant Sci 13. https://doi.org/10.3389/fpls.2022.984583

Daroodi Z, Taheri P, Tarighi S, Iranshahi M, Akaberi M (2024) Efficacy of ergosterol peroxide obtained from the endophytic fungus *Acrophialophora jodhpurensis* against *Rhizoctonia solani*. J Appl Microbiol 135(2):lxae031. https://doi.org/10.1093/jambio/lxae031. PMID: 38346851

Eskandari A, Danesh YR (2012) Molecular identification of *Glomus mosseae* from wheat rhizosphere in Damghan Region, Iran. J Appl Biol Sci 6(3):19–23

Esmaeilzadeh S, Zare-maivan H, Ghanati F (2006a) Study of mycorrhizal distribution of medicinal plants in Tandoureh National Park. Iran J Med Aromat Plants Res 21(4):489–504. https://doi.org/10.22092/ijmapr.2006.115031

Esmaeilzadeh S, Zare Mayvan H, Qanati F (2006b) Coexistence of vesicular and arbuscular mycorrhizae in medicinal plants of Tandore National Park. Iran Med Aromat Plants Res 21(4):489–504. https://doi.org/10.22092/ijmapr.2006.115031

Fattahi M, Mohammadkhani A, Shiran B, Baninasab B, Ravash R (2021) Evaluation of the symbiotic effect of mycorrhiza arbuscular on some pistachio rootstocks under salinity and drought conditions. J Crops Improvement 23(3):682–667

Frac M, Hannula SE, Bełka M, Jędryczka M (2018) Fungal biodiversity and their role in soil health. Front Microbiol 9:707. https://doi.org/10.3389/fmicb.2018.00707

Ghasemi S, Khodaei S, Karimi K, Tavakoli M, Pertot I, Arzanlou M (2019) Biodiversity study of endophytic fungi associated with two Quercus species in Iran. Forest Systems 28(1):e003–e003

Habibi M, zaefarian F, rejali F, bagheri N (2021) Isolation and identification of arbuscular mycorrhizal fungi of saffron symbiosis and investigation of their effect on yield and nutrient uptake under water interruption and different planting bed. J Crops Improvement 23(3):591–579. https://doi.org/10.22059/jci.2021.302406.2393

Hajieghrari B, Torabi-Giglou M, Mohammadi MR, Davari M (2008) Biological potential of some Iranian *Trichoderma* isolates in the control of soil borne plant pathogenic fungi. Afr J Biotechnol 7(8):967–972

Hardoim PR, van Overbeek LS, Berg G, Pirttilä AM, Compant S, Campisano A et al (2015) The hidden world within plants: ecological and evolutionary considerations for defining functioning of microbial endophytes. Microbiol Mol Biol Rev 79(3):293–320

Ibiang SR, Usami T, Sakamoto K, Ibiang YB (2023) Lettuce tolerance to verticillium wilt after inoculation with *Penicillium pinophilum* and *Rhizophagus intraradices*. Physiol Mol Plant Pathol 128:102171

Ilkhan L, Sedaghati E, Rahimzadeh Behzadi H, Hatami N (2023) The effect of arbuscular mycorrhizal fungi on growth and physiological indicators and absorption of some nutritional elements of chives (*Allium fistulosum*). Plant Process Function 12(56):239–262

Intana W, Kumla J, Suwannarach N, Sunpapao A (2023) Biological control potential of a soil fungus *Trichoderma asperellum* K1-02 against *Neoscytalidium dimidiatum* causing stem canker of dragon fruit. Physiol Mol Plant Pathol 128:102151

Jahromi M, Salehi Jozani G, Hosseini M, Khayyam Nekooee M, Akbari Vala S (2012) Isolation and identification of indigenous Arbuscular mycorrhizal fungi (AMF) symbiosis with crops in some Iran regions with drought conditions. Modares J Biotechnol 3(1):1–13

Kandil EK, Nofel MM, Abdelaziz AM, Mansour MM, Attia MS (2024) Effectual role of plant growth-promoting fungi and fosthiazate in controlling tomato root-knot nematode infection: *in vivo* and *in vitro* studies. Physiol Mol Plant Pathol 134:102463

Kashanian A, Panjekeh N, Taliei F (2021) Investigating diversity and spatial distribution of endophytic fungi in hazelnut (*Corylus avellana*) in its different habitats of Iran. J Microbial Biol 10(40):53–69

Kavehnia A, Samsampour D, Askari Seyahooei M (2023) Isolation and morphological and molecular identification of important endophyte fungi of Pistacia mutica in Hormozgan province and their effect on increasing salt stress tolerance of California Wonder 3 pepper seedlings. J Plant Prod Res 30(3):41–64

Legrifi I, Zhou J, Taoussi M, Radi M, Laasli SA, Al Figuigui J, Lazraq A, Chalot M, Lahlali R (2024) Genome sequencing and functional analysis of potential *Trichoderma* species for controlling *Pythium schmitthenneri*-induced root rot in olive trees. Physiol Mol Plant Pathol 134:102473. https://doi.org/10.1016/j.pmpp

Marrassini V, Ercoli L, Kuramae EE, Kowalchuk GA, Pellegrino E (2024) Arbuscular mycorrhizal fungi originated from soils with a fertility gradient highlight a strong intraspecies functional variability. Appl Soil Ecol 197:105344

Masoudi Khorasani F, Ganjeali A, Asili J, Cheniany M (2023) Beneficial effects of endophytic fungi inoculation on tanshinones and phenoliccompounds of *Salvia abrotanoides*. Iran J Basic Med Sci 26:408–413

Mehdinejad F, Zeynadini Riseh A, Sedaghati E, Alaei H, Moradi M (2021) Evaluation of the level of defense enzymes induced by antagonistic fungi against root knot nematode, *Meloidogyne javanica* in pistachio seedlings. J Plant Prot 35(1):25–37

Mirzaei J, Noorbakhsh N (2015) Isolation and identification of arbuscular mycorrhiza fungi in the roots and rhizosphere of *Amygdalus scoparia* Spach in Zagros forests (Case study: Kolm reservoir in Ilam province). Iran J Forest Range Prot Res 13(1):46–56

Modarresi Chahardehi A, Mousavi L, Bakhtkhah Ardeh Jani T, Rezaee Danesh Y, Ibrahim D (2014) Isolation and identification of dominant arbuscular mycorrhizal fungi in some trees in region of Kiasar. Iran J Plant Prot Sci 45(2):371–380

Nikmaram S, Sadravi M, Ghaderi Zefrehee M (2024) The impact of three arbuscular mycorrhizal fungi on wheat take-all disease caused by *Gaeumannomyces graminis* var. *tritici*. Plant Pathol Sci 13(1):104–112

Nooralvandi F, Alizadeh H, Sarami H, Salehi Jouzani G (2021) Morphological and molecular identification of *Fusarium* spp. as root endophytes of spontaneous barley (*Hordeum spontaneum*) and Turnip weed (*Rapistrum rugosum*). J Plant Prot 35(1):1–10

Osono T, Hirose D (2011) Colonization and lignin decomposition of pine needle litter by *Lophodermium pinastri*. For Pathol 41(2):156–162

Rajabi M, Azimi Moghadam MR, Azizi A, Soltani J (2022) Isolation and molecular identification of two rutin-producing endophytic fungi from Caper (*Capparis spinosa* L.). J Microbial Biol 11(44):169–180

Rana KL, Kour D, Sheikh I, Dhiman A (2019) Endophytic fungi: biodiversity, ecological significance, and potential industrial applications. In: Recent advancement in white biotechnology through fungi, vol 1: Diversity and Enzymes Perspectives. Springer Nature Press. https://doi.org/10.1007/978-3-030-10480-1-1

Rezaee Danesh Y (2013) Study on status of Arbuscular mycorrhizal fungi associated with barley in Damghan Region, Iran. Iran Plant Prot Res 26(4):437–449

Rezaee Danesh YR, Kariman K, Keskin N, Najafi S (2022) Characterization of arbuscular mycorrhizal fungal communities associated withvineyards in northwestern Iran. Turk J Agric For 46(3):271–279

Roy T, Pal N, Das N (2024) Regulation of the polyamine pool in plants: metabolic implications and stress mitigation, with emphasis on microbial influence. Physiol Mol Plant Pathol 132:102317

Sabet Jahromi M, Azizi A, Soltani J (2021) Diversity and spatiotemporal distribution of fungal endophytes associated with *Salvia multicaulis*. Curr Microbiol 78:1432–1447

Saderi SZ, Abrishamchi P, Ganjeali A, Radjabian T (2024) *Glomus mosseae* symbiosis improves disease resistance of *Melissa officinalis* to *Fusarium culmorum* through inducing antioxidant responses and plant hormones levels. Sci Hortic 338:113607

Sadravi M (2012) Arbuscular mycorrhizal fungi of alfalfa root in the Kohgiluyeh & Boierahmad province (SW Iran). Rostaniha 13(1):101–104

Samlikamnoed P, Anothai J, Chairin T (2023) Defense-related enzyme production in oil palm seedlings against basal stem rot pathogen Ganoderma boninense and its biological control by *Trichoderma asperellum*. Physiol Mol Plant Pathol 128:102154

Sedaghati E, Yazdanpanah M, Nadi M (2021) A review of taxonomic studies of arbuscular mycorrhizal fungi in Iran. Mycologia Iranica 8(2):15–30

Sellappan R, Thangavel K, Uthandi S (2023) Bioprotective potential of maize apoplastic fluid bacterium (*Bacillus amyloliquefaciens*) and arbuscular mycorrhizal fungi (*Glomus intraradices*) against Spodoptera frugiperda infestation in maize. Physiol Mol Plant Pathol 127:102050

Seyed Talebi FS, Safavi SA, Talaei-Hassanloui R, Bandani AR (2018) Study of virulence conidia germination types for some Beauveria bassiana isolates. Biol Control Pests Dis 7:65–73

Shahiri Tabarestani M (2022) Evaluation of antifungal effect of biodegradable nano encapsulated extract of *Trichoderma harzianum*. J Iran Plant Prot Res 36(2):183–195. (In Persian with English abstract). https://doi.org/10.22067/jpp.2022.75620.1081

Solhjouy-Fard S, Talaei-Hassanloui R, Maali-Amiri R, Sword GA (2023) Effect of endophytic *Beauveria bassiana* on growth traits of five rapeseed varieties *Brassica napus*. J Appl Res Plant Prot 12(1):87–94

Taheri P (2022) Crosstalk of nitro-oxidative stress and iron in plant immunity. Free Radic Biol Med 191(2022):137–149

Taheri P, Flaherty J (2022) Beneficial microorganisms in agriculture. Physiol Mol Plant Pathol 117:101777

Verbruggen E, Van Der Heijden MG, Weedon JT, Kowalchuk GA, Röling WF (2012) Community assembly, species richness and nestedness of arbuscular mycorrhizal fungi in agricultural soils. Mol Ecol 21(10):2341–2353

Wang D, Zhang W, Wang T, Li N, Mu H, Zhang J, Duan J (2015) Unveiling the mode of action of two antibacterial Tanshinone derivatives. Int J Mol Sci 16(8):17668–17681

Wang F, Hu T, Han Y, Zhao Z, Ren S, Pan Y, Wang D, Liu X (2023) Biological control of gentian spot blight by the combination of endophytic fungi fermentation broth and traditional Chinese medicine extractions. Physiol Mol Plant Pathol 128:102149

Wei H, He W, Mao X, Liao S, Wang Q, Wang Z, Tang M, Xu T, Chen H (2024) Arbuscular mycorrhizal fungi and exogenous Ca^{2+} application synergistically enhance salt and alkali resistance in perennial ryegrass through diverse adaptive strategies. Microbiol Res. https://doi.org/10.1016/j.micres.2024.127906

Wilpiszeski RL, Aufrecht JA, Retterer ST, Sullivan MB, Graham DE, Pierce EM, Zablocki OD, Palumbo AV, Elias DA (2019) Soil aggregate microbial communities: towards understanding microbiome interactions at biologically relevant scales. Appl Environ Microbiol 85. https://doi.org/10.1128/AEM.00324-19

Wu W, Cao Y, Cheng L, Wang L, Yu Q, Peng H, Zhou F, Liu H, Zhang Q (2022) Cryptotanshinone from *Salvia miltiorrhiza* inhibits the growth of tumors and enhances the efficacy of chemotherapy in a gastric cancer mouse model. Nat Prod Commun 17(10). https://doi.org/10.1177/1934578X221130874

Yao L, Ji C, Wang Z, Liu Y, Li W, Wang M, Yang X, He C, He X (2024) Enhancing Astragalus mongholicus performance through endophytic fungi improvement for fusarium wilt resistance. Physiol Mol Plant Pathol 136:102532

Yazdan Panah S, Ebrahim K, Pejman A, Mariah (2017) Morphological and molecular identification of arbuscular mycorrhizal fungi symbiotic with pistachio roots in Kerman province. Pistachio Sci Technol 2(4):11–29

Zamani SM, Matinizadeh M (2021) Investigating the mycorrhizal symbiosis of oak trees (*Quercus* sp.) in the forests of northern Iran. J Forest Wood Prod 74(3):331–343

Zanganeh S (2021) Report of some new species of *Acaulospora* genus in Iran. Rostaniha 22(1):67–74

Zangeneh S (2022) *Glomus nanolumen*, a new report from Iran. Rostaniha 23(2):309–310

Zare Z, Razavi SE, Rahnama K, Taheri A, Gherekhloo J (2023) Identification of fungal endophytes in aerobic rice cultivation of Golestan province. J Iran Plant Prot Res 37(2):105–117

Chapter 9
Research on Disease Suppressive Soils of Brazil

Rodrigo Pedrozo and Camila P. Nicolli

1 Introduction

Agriculture has long been the backbone of human civilization, providing vital resources that sustain populations and fuel economic development. However, this fundamental sector is increasingly threatened by a rising global population, shrinking availability of fertile land, and worsening environmental challenges such as soil degradation and climate change. Among these concerns, soilborne diseases are a major limitation to crop production, directly endangering food security worldwide. According to the Food and Agriculture Organization of the United Nations (FAO), plant diseases result in approximately US$220 billion in economic losses annually, with 20–40% of global crop yields affected by pests (FAO 2021). Furthermore, mounting evidence suggests that climate change influences the spread and severity of plant pathogens and pests, emphasizing the urgent need for sustainable, science-driven approaches to counteract these impacts.

A promising solution to this challenge lies in disease-suppressive soils, a naturally occurring phenomenon where certain soils exhibit a reduced occurrence of plant diseases due to their microbial composition. These soils harbor diverse microbial communities, including bacteria, fungi, archaea, and other microorganisms, which help protect plants by outcompeting pathogens, producing antagonistic compounds, and inducing plant systemic resistance (Raaijmakers and Mazzola 2016).

R. Pedrozo
USDA-ARS Dale Bumpers National Rice Research Center, Stuttgart, AR, USA

C. P. Nicolli (✉)
Department of Entomology and Plant Pathology, University of Arkansas System Division of Agriculture, Rice Research and Extension Center, Stuttgart, AR, USA
e-mail: cnicolli@uada.edu

S. Gnanamanickam, M. Höfte (eds.), *Disease Suppressive Tropical Soils*,
https://doi.org/10.1007/978-3-032-08100-1_9

The native soil microbiome plays a crucial role in plant health and resilience (Singh et al. 2020). Additionally, the structure, function, and diversity of bulk and rhizosphere soil microbiomes are known to be shaped by interactions with soilborne pathogens (Poudel et al. 2016). Since the rhizosphere microbiome serves as a plant's first line of defense against soilborne diseases, the interactions in this region significantly impact disease incidence and severity (Berendsen et al. 2012; Chiaramonte et al. 2021; Yang et al. 2017).

The importance of disease-suppressive soils is particularly evident in tropical agricultural regions like Brazil. In these areas, while crop yields are high, the prevalence of plant diseases is also significant. Brazil's tropical climate allows for highly productive farming, but it also creates conditions that favor the survival and spread of soilborne pathogens (Batista et al. 2024; Ghini et al. 2011; Wen et al. 2017). This unique situation offers an opportunity to explore the potential of naturally suppressive soils as a strategy to reduce crop losses and promote sustainable agricultural practices.

For the 2024/2025 growing season, Brazil is on track to achieve its largest grain harvest in history, projected to reach 322.3 million tons, an 8.2% increase from the previous season. The main contributors to this surge are soybeans and corn, with soybean production expected to reach 166.3 million tons and total corn production estimated at 119.6 million tons (CONAB 2024). Notably, Brazil ranks as the world's largest producer of soybeans, surpassing the United States and Argentina, which hold the second and third positions, respectively. This dominance in soybean production is attributed to Brazil's extensive farmland, favorable growing conditions, and technological advancements in agricultural practices. However, such intensive production systems pose challenges, particularly concerning soil health and disease management.

Conversely, suppressive soils exemplify natural biological disease control, where native antagonistic microorganisms play a crucial role in limiting pathogen proliferation. Discussions regarding plant protection and sustainable agriculture have increasingly focused on concepts like soil health and soil quality, which align with broader movements such as ecological intensification, regenerative agriculture, microbiome-based bio-inputs (including all its derivatives), and Agriculture 5.0 (Table 9.1). While these terms are not interchangeable, they share common principles and contribute to a more holistic approach to sustainable farming.

The terminology and methods used to study cultivated soils may vary depending on research priorities, but the central goal remains the same: to develop long-term solutions for improving soil function and plant resilience (Eilenberg et al. 2001). This chapter will explore disease-suppressive soils as a key strategy for managing root diseases, with a specific focus on Brazil's soybean production system.

Table 9.1 Glossary of essential concepts in soil suppressiveness and sustainable agriculture

Concept	Definition	Additional Details	Scientific References
Suppressive Soil	Soils where the pathogen does not establish or persist; establishes but causes little or no damage; or where disease severity/incidence remains low and becomes insignificant despite the presence of the pathogen, a susceptible host, and favorable climatic conditions.	Suppressiveness is linked to microbial diversity, organic matter content, and ecological balance in the rhizosphere. Strategies such as crop rotation, organic amendments, and microbial inoculants help enhance suppressiveness.	Baker and Cook (1974)
Soil Quality	The ability of a specific type of soil to function within ecosystem limits—natural or managed—to sustain plant and animal productivity, maintain or improve water and air quality, and support human health and habitation.	Soil quality assessments integrate physical (structure, water retention), chemical (pH, nutrients), and biological (microbial biomass, enzymatic activity) indicators.	Karlen et al. (1997)
Soil Health	The continuous ability of soil to function as a vital system, within ecosystem and land-use limits, to sustain biological productivity, maintain air and water quality, and promote plant, animal, and human health.	Soil health is linked to ecosystem services, including carbon sequestration, disease suppression, and nutrient cycling. Microbial communities play a crucial role in maintaining soil resilience.	Doran et al. (1996)
Bio-fertilizers	Live microorganisms that improve nutrient availability by fixing nitrogen, solubilizing phosphorus, or enhancing nutrient uptake.	Examples include rhizobia, *Azospirillum*, *Azotobacter*, and phosphate-solubilizing bacteria. Biofertilizers play a key role in sustainable nutrient management and reduce dependence on chemical fertilizers.	Vessey (2003)
Bio-pesticides	Biological agents, including microorganisms, plant extracts, and naturally occurring compounds, that control pests, pathogens, or weeds.	Includes bacterial-based (e.g., *Bacillus thuringiensis*), fungal-based (e.g., *Beauveria bassiana*), viral-based (e.g., baculoviruses), and plant-extract-based biopesticides.	Glare and Nollet (2023)
Ecological Intensification	A system that seeks to maximize productivity without compromising the system's long-term productive capacity.	Focuses on optimizing input efficiency through precision agriculture, integrated pest management (IPM), and digital monitoring of soil and crop health.	FAO (2009)

(continued)

Table 9.1 (continued)

Concept	Definition	Additional Details	Scientific References
Regenerative Agriculture	A system redesign aimed at restoring resource bases through natural ecological services, with the primary goal of improving soil health or restoring highly degraded soil while minimizing or even creating positive environmental impacts.	Involves practices such as no-till, cover crops, crop diversification, and livestock integration. Enhances biodiversity, promotes carbon sequestration, and reduces dependency on synthetic inputs.	Rhodes (2017)
Microbiome-Based Bio-Inputs	A broad category of biological inputs derived from soil microbiomes, including probiotics, prebiotics, biostimulants, biofertilizers, and biopesticides, designed to improve soil suppressiveness, plant resilience, and nutrient efficiency.	Advances in metagenomics, synthetic biology, and AI-driven microbial consortia selection are driving the next generation of biological soil amendments for sustainable disease management.	Foo et al. (2017), Ke et al. (2021)
Bio-stimulants	Substances or microorganisms that enhance plant growth, nutrient uptake, and stress tolerance without directly providing nutrients or controlling pests.	Includes microbial and non-microbial biostimulants such as humic acids, seaweed extracts, amino acids, mycorrhizal fungi, and rhizobacteria. They enhance root development, nutrient efficiency, and stress resistance.	du Jardin (2015)
Pro-biotics/ Bioinoculants (for Plants)	Beneficial microorganisms applied to plants and soil to enhance microbial diversity, plant resilience, and disease resistance.	Includes probiotic consortia of beneficial bacteria and fungi that colonize the rhizosphere, promote plant growth, and suppress pathogens. Often used in biocontrol and plant stress adaptation. Used in seed treatments, soil applications, and foliar sprays. Contains species such as Trichoderma, Pseudomonas, and Bacillus, which act as growth promoters and disease suppressors.	Backer et al. (2018), Bashan et al. (2014)
Pre-biotics	Non-living compounds that stimulate the growth and activity of beneficial soil microorganisms, improving plant health.	Includes organic acids, carbohydrates, and plant-based compounds that promote beneficial microbes in the soil microbiome, enhancing suppressive soil properties.	Berg et al. (2020a)

(continued)

Table 9.1 (continued)

Concept	Definition	Additional Details	Scientific References
Agriculture 5.0	The next phase of agricultural innovation integrating AI, IoT, robotics, nanotechnology, and microbiome-engineered solutions to enhance sustainability and productivity.	Agriculture 5.0 involves real-time soil microbiome analysis, AI-driven crop health monitoring, autonomous drones, and bio-digital twins that simulate plant-soil-microbe interactions. Advances in omics technologies (metabolomics, transcriptomics) are helping optimize microbiome functions for disease suppression and nutrient efficiency.	Abdul Razak et al. (2024)

[a]These concepts used in soil research can vary depending on scientific priorities, disciplinary perspectives, and technological advancements. By integrating ecological principles, microbiome research, and innovative agricultural practices, long-term solutions can be created to maintain soil health, optimize nutrient cycling, and mitigate the impact of biotic and abiotic stresses on crops

2 Managing Soilborne Pathogens Trough Suppressive Soils in Brazil

Brazil has become one of the world's leading agricultural producers, experiencing a 206% rise in crop productivity and a 394% increase in grain production over the past four decades (CONAB 2020). This impressive progress has been driven by advancements in plant breeding, conservationist soil management, mechanization, government-supported agricultural policies, and investments in cutting-edge farming technologies (CONAB 2023). The adoption of conservationist farming practices has expanded considerably, with a 70% increase worldwide from 2009 to 2016. In Brazil, these methods are now applied to over 32 million hectares, accounting for more than half of the 60 million hectares of cultivated land. This shift has been gradual, evolving alongside improvements in grain production management strategies (Kassam et al. 2019; Mendes et al. 2018). Among the many benefits of conservation-based agriculture, one of the most significant is its role in fostering suppressive soil conditions, which help reduce the incidence and severity of soilborne diseases, a central theme of this discussion.

The mechanisms by which suppressive soils inhibit pathogen growth are complex and involve a delicate balance of biological, chemical, and physical factors. Brazil's tropical climate and intensive agricultural practices create favorable conditions for soilborne diseases, making the country an excellent case study for understanding how suppressive soils function (Castro Lopes et al. 2018; Ghini et al. 2011). One key factor influencing pathogen survival in the soil is climate, which introduces seasonal variability in disease prevalence. Environmental parameters such as temperature, rainfall, and soil moisture significantly influence root diseases and the effectiveness of control strategies (de Freitas and Landers 2014; Raza et al. 2017).

The activity of pathogens is influenced by the availability of organic matter, as many microorganisms rely on decomposing plant material at some stage in their lifecycle. The presence of host plants directly affects the levels of pathogen inoculum, with changes in crop residues and root material impacting the survival and spread of disease-causing organisms. These factors can either promote or inhibit the establishment of pathogens, depending on how they interact within the soil ecosystem. For example, a widely adopted strategy for controlling soilborne diseases includes crop rotation, the incorporation of organic matter, and soil disinfestation. A crucial measure for evaluating the effectiveness of these methods is inoculum density, as the probability of plant infections is strongly linked to the number of pathogen propagules present in THE soil (Andrade Botelho et al. n.d.; Michereff et al. 2005). The relationship between disease incidence and pathogen density has been thoroughly examined in various systems, including, *Macrophomina phaseolina*, *Rhizoctonia solani*, *Fusarium oxysporum*, and *Sclerotinia sclerotiorum*. Since soilborne pathogens have limited natural mobility, monitoring their population densities over time offers valuable insights for assessing disease risk (Geraldine et al. 2013; Meyer et al. n.d.; Srivastava et al. 1996; de Toledo-Souza et al. 2008, 2012).

Some reports suggest that monoculture farming can, under specific circumstances, reduce soil pathogen populations. However, most of these findings come from temperate regions, where annual cropping cycles naturally limit pathogen persistence (Westphal 2005). In contrast, Brazil's tropical and subtropical climates experience continuous cultivation throughout the year, leading to significantly different pathogen behaviors. The implementation of no-till crop rotation systems (SPD) has played a crucial role in reducing soil inoculum loads and managing root diseases (Almeida et al. 2001; de Toledo-Souza et al. 2008). However, simply lowering inoculum density does not always suffice to control diseases effectively, especially in short-rotation systems where pathogens with broad host ranges, such as *Sclerotinia sclerotiorum*, *Macrophomina phaseolina*, and *Rhizoctonia solani*, continue to pose challenges (Blanco et al. 2018; Martins Alves et al. 2021; Rodrigues et al. 1998; Srivastava et al. 1996).

Research conducted in Brazilian soybean fields has shown that the prevalence of those diseases, charcoal rot, white mold, and root rot increase over time, mainly due to the persistence of the pathogens in fields with a long history of soybean monoculture (Almeida et al. 2001; Geraldine et al. 2013). An ongoing question in root disease epidemiology is the relationship between inoculum density and disease severity (Campbell 1994; Madden et al. 2007; Parker and Gilbert 2018). Even when the population of pathogen propagules remains constant, their ability to infect plants can fluctuate due to factors such as nutritional availability, environmental conditions, and microbial competition (Madden et al. 2007). The survival and spread of pathogens in soil are regulated by density-dependent processes that are influenced by resource availability, competition with other microbes, and the carrying capacity of the environment (Westphal 2005). Additionally, beneficial soil microorganisms can affect pathogen survival since microbial interactions often suppress disease development through resource competition and antagonistic activity (Parker and Gilbert 2018). While inoculum density is a crucial factor in determining disease risk, it does not completely explain why certain pathogens thrive in specific conditions while remaining dormant in others.

This chapter does not concentrate on the widespread application of antagonistic microorganisms for biological control, such as *Trichoderma harzianum* and *Pseudomonas fluorescens*. However, it is important to highlight that these bioagents can enhance disease suppression by decreasing initial pathogen inoculum levels, which is especially beneficial in managing *M. phaseolina*. Biocontrol agents that are formulated with high concentrations of viable conidia or bacterial cells generally have a minimal long-term impact on soil microbial diversity (de Azevedo Silva et al. 2021; Srivastava et al. 1996; Topalović et al. 2020); making them compatible with conservation-based disease control strategies.

The mere presence of pathogens in the soil does not necessarily lead to active disease development, as competition for nutrients, infection sites, and environmental conditions can significantly limit pathogen activity. In such cases, fungistasis, the ability of soil to inhibit fungal spore germination and suppress hyphal growth, can occur even when temperature and moisture levels are optimal. It has been reported that this process is primarily driven by two mechanisms: the release of volatile antifungal compounds that inhibit pathogen activity and intense microbial competition for organic nutrients, which creates an environment that is unfavorable for pathogen survival (Bonanomi et al. 2017; Watson and Ford 1972). Studies conducted in Brazil have shown that *Macrophomina phaseolina* can survive in plant residues, particularly in buried root and stem fragments. This ability allows the pathogen to remain viable between cropping cycles (Almeida et al. 2001). Understanding the interactions between this pathogen and the soil is crucial for developing disease-suppressive soils, which are increasingly needed to meet the growing demand for food and fiber.

One promising strategy to enhance soil suppressiveness is to stimulate native microbial communities, such as arbuscular mycorrhizal fungi or plant growth promoting rhizobacteria (PGPR) (Bashan et al. 2014; de Souza et al. 2015; Vessey 2003). These fungi and bacterial communities not only improve plant nutrient uptake but also reduce the severity of charcoal rot in soybeans and suppress *Fusarium solani* propagules in bean roots (de Azevedo Silva et al. 2021; Hassan Dar et al. 1997; de Toledo-Souza et al. 2012). Since soilborne pathogens exist within complex microbial ecosystems, fostering beneficial microbial interactions appears to be an effective long-term approach to enhancing suppressive soils in Brazilian agricultural systems (De Corato 2020; Postma et al. 2008).

3 Monitoring Disease Soil Suppressiveness

There is no single, universally accepted method for assessing soil suppressiveness. However, key parameters such as disease prevalence, incidence, and severity are essential for evaluating crop management outcomes and constraints (Campbell 1994; Chiaramonte et al. 2021). Additionally, various soil attributes have been studied as potential indicators of suppressiveness, soil quality, and overall soil health. These indicators help monitor land use patterns, degradation processes, and the impacts of different agricultural systems.

Soil quality assessments typically categorize these indicators into physical, chemical, and biological aspects. Among these, biological indicators are particularly valuable because they focus on the composition, activity, and diversity of soil microorganisms. Since microbial communities are highly responsive to environmental disturbances, they can serve as early-warning indicators of changes in soil conditions. One commonly used metric in soil health assessments is microbial biomass carbon, which often responds more quickly to shifts in soil environments compared to physical or chemical attributes (Hungria et al. 2009).

The microbial biomass fraction within soil constitutes the dynamic living component of organic matter, comprising microorganisms that are typically smaller than 5×10^3 μm^3. This diverse assemblage includes various taxonomic groups such as Archaea, Bacteria, Fungi, Protozoa, and specific microscopic representatives from the Animalia kingdom like nematodes (Moreira and Siqueira 2006; Pietikäinen and Fritze 1995).

An equally significant metric for understanding soil health is soil basal respiration, which quantifies the overall metabolic activity within the soil ecosystem, reflecting the total carbon dioxide (CO_2) emissions resulting from microbial metabolism (Evangelista Da Silva et al. n.d.; Pietikäinen and Fritze 1995). Fungi and bacteria are paramount in the decomposition of organic matter, facilitating the breakdown of complex organic compounds into simpler forms. Consequently, the volume of CO_2 released during these processes is predominantly associated with the activity of heterotrophic microbial communities, which play a critical role in the cycling of carbon through the soil.

Under natural conditions, factors such as temperature and moisture levels exert substantial influence on soil respiration rates. Soil temperature enhances enzymatic activity among microbes, often leading to increased respiration rates during warmer periods. Conversely, moisture availability is crucial, as it affects microbial health and metabolic function. While elevated CO_2 emissions can signify intense microbial activity and rapid turnover of organic matter, the interpretation of these rates is nuanced and complex. Prolonged periods of excessive respiratory activity may indicate not only a vigorous microbial metabolism but also the potential for mineralization of soil carbon reserves, which can result in carbon depletion. This depletion may impair the soil's ability to suppress pathogens and ultimately diminish its resilience to environmental stresses. To enhance our understanding of microbial efficiency, Anderson and Domsch (1993) introduced the innovative metabolic quotient

(qCO_2), a pivotal metric that compares soil respiration to microbial biomass over time. Unlike traditional standalone measurements of biomass or respiration, qCO_2 offers a comprehensive perspective into the intricate dynamics of microbial activity, accounting for both environmental factors and anthropogenic impacts. This index serves as a vital tool for evaluating microbial stress, driven by challenges such as nutrient depletion, pollution, or unfavorable soil pH levels. A lower qCO_2 value signifies a remarkable shift: as microbial populations increase their resource utilization efficiency, less carbon is released as CO_2, indicating that more carbon is being stored within microbial biomass. This insight is crucial for understanding carbon retention within ecosystems (Anderson and Domsch 1993; Evangelista Da Silva et al. n.d.; Hungria et al. 2009).

The rise of precision agriculture technologies has dramatically changed the way soil health is monitored. Cutting-edge tools like remote sensing and soil health monitoring systems now facilitate real-time evaluations of microbial activity, nutrient levels, and pathogen presence. Ongoing research initiatives in Brazil and around the globe are making strides to incorporate biological indicators into commercial soil testing procedures. Among the promising methods, soil enzyme activity has emerged as an influential approach to indirectly assess microbial function and swiftly detect changes in soil management practices. A significant advancement in this domain was achieved by Embrapa with the development of BioAs technology, which quantifies the activity of arylsulfatase and β-glucosidase as key indicators of soil biological health. These enzymes are closely correlated with microbial biomass carbon and are relatively easy to measure (Mendes et al. 2018; Sobucki et al. 2021). However, a critical research gap remains: the direct connections between these enzymatic markers and the prevalence, severity, or resultant crop yield losses from diseases have yet to be established (Janvier et al. 2007). This highlights the urgent need for further investigations into their potential as indicators of disease suppression. With dedicated research efforts, the agricultural community can empower farmers with enhanced insights into their soil conditions, paving the way for more effective and sustainable disease control strategies.

Additionally, BioAS aids in calculating Soil Quality Indexes (IQS), which integrate biological and chemical aspects of soil properties (IQSFertbio) to deliver a comprehensive evaluation of soil health (Silva et al. 2023). In 2020, the Brazilian government initiated the National Bioinputs Program, intended to enhance and develop the bioinputs sector (*MAPA – Ministério Da Agricultura e Pecuária. Principais Marcos Históricos Dos 160 Anos Do MAPA*, 2020). This program offers farmers access to biologically-based agricultural inputs, such as fertilizers, pest management solutions, and post-harvest treatments, while also improving the cost-effectiveness and efficiency of biostimulants and organic amendments.

To truly grasp the complexities of soil microbial diversity and its vital functions, we must harness cutting-edge molecular techniques, particularly in the realm of microbial genomics. The emergence of next-generation sequencing (NGS) technologies, coupled with innovative bioinformatics tools, has sparked an exciting wave of research aimed at mapping microbial communities across varying soil environments (Bouchez et al. 2016; Mendes et al. 2011). Rather than simply cataloging microbial compositions, scientists are now delving deeper to uncover the functional traits and intricate interactions among these tiny organisms. This burgeoning field of microbial ecology holds the promise of revealing new bioindicators that could either supplement or replace our current measures of soil health (Chiaramonte et al. 2021). Gaining a richer understanding of soil microbial dynamics is crucial for advancing sustainable agricultural practices and tackling the pressing issue of global soil degradation (Bünemann et al. 2018; Hartmann et al. 2015).

One of the most fascinating developments is the increasing application of molecular methods that allow for the differentiation of microbial communities across various soil management systems. This research enables scientists to visualize microbial structures and gauge levels of biodiversity. A prime example of this innovative work can be seen in the study by Gardener and Weller (2001), which revealed significant changes in wheat rhizosphere microbial populations. They discovered a remarkable increase in *Pseudomonas fluorescens* strains that produce 2,4-diacetylphloroglucinol (2,4-DAPG), a powerful compound known for its antifungal properties against *Gaeumannomyces graminis* var. *tritici*, the culprit behind take-all disease. This case remains one of the most extensively researched instances of disease suppression in monoculture systems. Interestingly, different 2,4-DAPG-producing strains exhibit unique colonization patterns, showcasing host-specific preferences. For instance, genotype D strains are notably more frequent in association with wheat and pea plants (McSpadden Gardener and Weller 2001).

Looking ahead, efforts to enhance soil health databases and refine data analysis methodologies are set to improve indices such as the Soil Management Assessment Framework (SMAF) and the Comprehensive Assessment of Soil Health (CASH). These frameworks are expected to evolve, integrating new classes of indicators like molecular diagnostics, comprehensive soil profile data, and improved sensitivity assessments. As SMAF and CASH continue to advance, they will empower farmers with precise insights into soil constraints, allowing for more informed soil and crop management decisions that foster sustainability and promote robust, long-term soil health (Karlen et al. 1997, 2019; Lehman et al. 2015).

One of the most groundbreaking innovations in soil analysis is the use of spectroscopic methods. Near-infrared spectroscopy and remote sensing technologies have emerged as powerful, cost-effective solutions for accurately assessing the chemical, physical, and biological properties of soil (Chabrillat et al. 2019; Gandariasbeitia et al. 2017; Paz-Kagan et al. 2014). These advanced techniques have been effectively employed to estimate key indicators such as organic carbon content, total nitrogen levels, β-glucosidase activity, microbial biomass carbon, particulate organic matter, and soil respiration (Chang et al. 2001; Paustian et al. 2019; Pietikäinen and Fritze 1995; Veum et al. 2015).

In a significant step forward for large-scale soil health monitoring, the Brazilian Soil Spectral Service (BraSpecS) has been launched as an innovative online tool designed to predict critical soil attributes, including texture, organic carbon content, and various nutrient levels (Demattê et al. 2022). By seamlessly integrating traditional phytopathology with advanced molecular tools and cutting-edge digital technologies, this initiative is set to revolutionize soil research, paving the way for more effective and sustainable strategies for disease suppression. This approach not only enhances our understanding of soil health but also fosters a more resilient agricultural ecosystem.

4 Innovations in Soil Management

The increasing knowledge of disease-suppressive soils has sparked a wave of advancements in soil management practices, particularly in the use of biostimulants, including their derivatives such as probiotics and prebiotics for plants, alongside integrated sustainable farming techniques. Biostimulants are defined as substances that, when applied to seeds, plants, or soil, stimulate natural biological processes to improve nutrient uptake, nutrient efficiency, tolerance to biotic and abiotic stresses, and overall crop productivity and quality, independent of their nutrient content or microbial composition (Albrecht 2019). These emerging technologies have revolutionized how researchers and farmers tackle soil health and plant resilience, particularly in Brazil's intensively cultivated yet disease-prone landscapes (Bergez et al. n.d.).

A particularly noteworthy advancement has been the development of probiotics, also called bioinoculants, which consist of formulations containing living beneficial microorganisms that play a role in soil health and plant disease prevention. With advancements in biotechnology, it has become possible to isolate, identify, and develop bioinoculants that specifically target soilborne pathogens. For instance, formulations based on *Bacillus subtilis* and *Trichoderma harzianum* have been commercialized to combat soybean root rot and white mold in Brazil (Chagas et al. 2018; de Azevedo Silva et al. 2021; Hassan Dar et al. 1997; Nascimento et al. 2022). These bioinoculants introduce high concentrations of beneficial microbes into the rhizosphere, where they effectively inhibit pathogens by employing mechanisms such as antibiotic production, competition for nutrients, and induction of systemic resistance (Geraldine et al. 2013; Raaijmakers and Mazzola 2016).

Recent research has led to the introduction of microbial consortia, which integrate multiple bacterial and fungal strains to generate synergistic effects that boost disease suppression while improving soil resilience to environmental challenges. Trials conducted in Brazilian soybean fields have demonstrated that these multispecies microbial inoculants, such as combinations of *Bradyrhizobium* spp. and *Azospirillum brasilense*, perform significantly better than single-strain formulations, leading to notable improvements in soil health, plant vigor, and resistance to stress conditions (Nunes et al. 2024). Historically, soybeans have played a pivotal role in the advancement of bio-input applications in Brazilian agriculture. The early successes of biological nitrogen fixation with *Bradyrhizobium* spp. and the use of *Bacillus thuringiensis* for soybean caterpillar control marked the first major milestones in the adoption of biological solutions in Brazilian farming (Glare and Nollet 2023; Hungria et al. 2009).

The relationship between soybean cultivation and the expansion of bio-inputs in Brazil is evident both in historical developments and in contemporary farming practices (Mendes et al. 2014). The increasing demand for environmentally sustainable agricultural techniques has encouraged the widespread use of beneficial microbial species, particularly Bacillus-based bio-inputs, which have been found to not only aid in pest and disease control but also contribute to plant growth stimulation. The extensive use of *Bacillus*-based technologies in Brazil highlights the alignment between sustainable agricultural practices and continuous innovation in soybean farming (Civardi et al. 2019; Goulet 2021). Studies indicate that applying *Bacillus* strains alone or in combination with *Trichoderma* has resulted in substantial improvements in biomass production in various crops, including soybean, rice (*Oryza sativa* L.), cowpea (*Vigna unguiculata* (L.) Walp), and maize. These findings underscore the role of *Bacillus* as a plant growth promoter, particularly in cowpea and maize, where significant increases in root and shoot biomass have been recorded (Chagas et al. 2018). Further supporting this, research by Nain et al. (2012) confirmed the findings of Chagas et al. (2018), demonstrating that the inoculation of cowpea seeds with *Bacillus* sp. strain RM-2 resulted in enhanced germination, increased leaf area, longer shoot and root systems, higher pod counts, and greater grain yield per plant (Minaxi et al. 2012).

Alongside probiotics, prebiotics have emerged as another promising category of biostimulants gaining traction in sustainable agriculture. Unlike probiotic-based bioinoculants, prebiotics are non-living organic and inorganic compounds that stimulate beneficial soil microbes to enhance soil health. These products typically include humic and fulvic acids, protein hydrolysates (from plant or animal sources), seaweed extracts, chitosan, biopolymers, composted materials (such as manure or sewage sludge), and essential minerals such as silicon, iron (Fe), manganese (Mn), and zinc (Zn) (Bulgari et al. 2015; Calvo et al. 2014). Prebiotic biostimulants are commonly administered via soil drenching, where they improve soil structure, fertility, and microbial diversity (du Jardin 2015; Yakhin et al. 2017). Over the past decade, research interest in soil prebiotics has surged, coinciding with a rapid increase in commercial biostimulant products worldwide (Backer et al. 2018; Ricci et al. 2019). The growing adoption of prebiotics has helped reduce reliance on synthetic fertilizers, pesticides, and herbicides, minimizing their negative impact on soil biodiversity and ecosystem balance (Rouphael and Colla 2020). The application of biostimulants has proven highly effective in improving plant vigor, productivity, and soil suppressiveness (Berg et al. 2020b; Colla et al. 2017; Nosheen et al. 2021).

Brazil has also seen the development and commercial success of various biostimulants, with Microgeo® standing out as a prime example (de Almeida et al. 2018; Suárez et al. 2020; Rangel da Silva Gama et al. 2014). Microgeo® has been a well-established biostimulant in South America for over two decades, with widespread adoption across diverse agricultural systems. This product operates through a continuous liquid composting process, wherein locally adapted microorganisms are cultivated on-site in biofactories, using an organomineral matrix to optimize their function (Da Cunha Leme Filho et al. 2021; Posso Suárez et al. 2020).

According to the 2024 Dunham Trimmer annual biological market report, the global biological input industry is currently valued at US$11 billion, with projections estimating it will reach US$18 billion by 2029. Among global markets, Brazil is experiencing the fastest growth, with its biological products market doubling in size every 2 years. Importantly, Brazil is the only nation that has widely implemented biocontrol solutions in large-scale row crops, highlighting its leadership in biological-based agricultural solutions.

Beyond microbial inoculants, the integration of soil disease suppression strategies with national agricultural policies has played a crucial role in promoting sustainable farming in Brazil. Policymakers have increasingly emphasized soil health as a national priority, incentivizing the use of organic amendments, diversified crop rotations, and biological control agents. For instance, subsidies for the recycling of agricultural waste, such as sugarcane bagasse and coffee husks, have encouraged their application as organic fertilizers, enriched microbial diversity and enhanced soil suppressiveness (De Corato 2020; Lobo et al. n.d.). These organic amendments act as substrates for beneficial microbes, fostering their proliferation and strengthening natural pathogen suppression mechanisms. Furthermore, the Brazilian government has promoted educational initiatives for farmers, offering hands-on training, demonstrations, and technical support to facilitate the transition toward biologically based agricultural systems (Freitas et al. n.d.; Singh et al. 2020).

Collectively, these innovations signal a major transformation in soil management, shifting away from chemical-intensive practices toward sustainable, biologically driven solutions. By merging traditional agronomic knowledge with cutting-edge scientific advancements, Brazil is establishing itself as a global leader in sustainable agriculture. The widespread adoption of bioinoculants, organic amendments, and precision agricultural tools is not only addressing present-day challenges but is also laying the groundwork for a more resilient and productive agricultural future.

5 Conclusion

The study and practical application of disease-suppressive soils represent a significant breakthrough in sustainable agriculture, offering solutions to major challenges such as soilborne diseases, environmental degradation, and the growing demand for food production. Brazil, with its vast and varied agricultural ecosystems, has positioned itself as a global leader in utilizing these soils to improve crop resilience and overall productivity. By harnessing natural microbial communities and incorporating them into modern farming systems, these soils present an innovative way to reduce dependency on chemical inputs while fostering a healthier environment.

The effectiveness of disease-suppressive soils is influenced by the adoption of integrated soil management techniques, such as crop rotation, soil tillage, incorporation of fertilizers or manure, pesticide usage, and irrigation methods. Each of these factors plays a role in shaping soil microbial dynamics, ultimately impacting its suppressive capacity (Garbeva et al. 2004). Rather than viewing these techniques in isolation, a comprehensive approach that considers the entire agricultural system is key to achieving optimal soil suppressiveness. Several biological mechanisms contribute to this phenomenon, including microbial antagonism, in which beneficial microbes generate enzymes and secondary metabolites that directly counteract pathogens; competition for nutrients and space, demonstrated by siderophore-producing Pseudomonas, which limits pathogen access to essential resources; induced systemic resistance, where beneficial microbes activate plant immune responses, making crops more resilient to infections; and organic matter interactions, which release bioactive substances that not only inhibit harmful microbes but also stimulate beneficial microbial populations. Together, these processes highlight the transformational impact of disease-suppressive soils in Brazilian agriculture, offering sustainable and effective crop protection strategies that align with the goals of reducing chemical reliance and enhancing soil health.

Advancements in soil management, particularly in the development of biostimulants and microbial consortia, have accelerated the adoption of disease-suppressive strategies. Programs aimed at promoting organic amendments, rotational cropping, and precision agriculture have provided Brazilian farmers with tools to maximize soil suppressiveness, ensuring long-term soil fertility and crop productivity. These approaches not only improve microbial diversity and biological activity but also play a crucial role in carbon sequestration and biodiversity preservation (Paustian et al. 2019).

In addition, cutting-edge technologies, such as metagenomics and remote sensing, have allowed for a deeper understanding of soil microbial communities, facilitating the design of more targeted and effective soil health interventions (Mendes et al. 2011; Raaijmakers and Mazzola 2016). However, despite these advancements, several obstacles remain in fully leveraging the potential of disease-suppressive soils. Challenges such as climate variability, land degradation, and the need for large-scale implementation continue to be significant barriers (Velásquez et al. 2018). Addressing these issues will require collaborative engagement between scientists, policymakers, and farmers, as well as increased investments in education, infrastructure, and regulatory support to ensure that smallholder farmers, who form the foundation of Brazil's agricultural sector, can also benefit from these advances.

Moving forward, the success of disease-suppressive soils will depend on striking a balance between productivity and environmental responsibility. Brazil's leadership in this area serves as an exemplary model for other nations, demonstrating the importance of integrating traditional farming wisdom with modern scientific research. By continuing to support agricultural research, technological innovation, and policy advancements, Brazil can reinforce its position as a global reference in sustainable agriculture, making critical contributions to food security and environmental conservation worldwide. Ultimately, disease-suppressive soils represent a transformative tool in reshaping agriculture into a more sustainable, resilient, and efficient system. The Brazilian experience showcases how scientific progress, coupled with collaborative efforts, can help tackle some of the most urgent agricultural challenges. As the world faces growing concerns related to climate change and food insecurity ("Pathogens, Precipitation and Produce Prices," 2021), Brazil's advancements in soil suppressiveness research and application offer valuable lessons and a roadmap toward a more sustainable agricultural future.

References

Abdul Razak SF, Yogarayan S, Sayeed MS, Mohd Derafi MIF (2024) Agriculture 5.0 and explainable AI for smart agriculture: a scoping review. Emerg Sci J 8(2):744–760. https://doi.org/10.28991/ESJ-2024-08-02-024

Albrecht U (2019) Plant biostimulants: definition and overview of categories and effects. EDIS 2019(3). https://doi.org/10.32473/edis-hs1330-2019

Almeida ÁMR, Saraiva OF, Farias JRB, Gaudêncio CA, Torres E (2001) Survival of pathogens on soybean debris under no-tillage and conventional tillage systems. Pesq Agrop Brasileira 36(10):1231–1238. https://doi.org/10.1590/S0100-204X2001001000003

Anderson T, Domsch K (1993) The metabolic quotient for CO2 (qCO2) as a specific activity parameter to assess the effects of environmental conditions, such as ph, on the microbial biomass of forest soils. Soil Biol Biochem 25(3):393–395. https://doi.org/10.1016/0038-0717(93)90140-7

Andrade Botelho S, Rava CA, Leandro WM, Luis J, Costa S (n.d.) Supressividade natural de solos da Região Centro-Oeste A Rhizoctonia solani KÜHN 1. Pesquisa Agropecuária Tropical 31(2)

Backer R, Rokem JS, Ilangumaran G, Lamont J, Praslickova D, Ricci E, Subramanian S, Smith DL (2018) Plant growth-promoting rhizobacteria: context, mechanisms of action, and roadmap to commercialization of biostimulants for sustainable agriculture. Front Plant Sci 9:1473. https://doi.org/10.3389/fpls.2018.01473

Baker R, Cook J (1974) Biological control of plant pathogens San Francisco. W. H. Freeman

Bashan Y, de-Bashan LE, Prabhu SR, Hernandez J-P (2014) Advances in plant growth-promoting bacterial inoculant technology: formulations and practical perspectives (1998–2013). Plant Soil 378(1–2):1–33. https://doi.org/10.1007/s11104-013-1956-x

Batista BD, Wang J, Liu H, Kaur S, Macdonald CA, Qiu Z, Trivedi P, Delgado-Baquerizo M, Xiong C, Liang J, Bange M, Singh BK (2024) Biotic and abiotic responses to soilborne pathogens and environmental predictors of soil health. Soil Biol Biochem 189:109246. https://doi.org/10.1016/j.soilbio.2023.109246

Berendsen RL, Pieterse CMJ, Bakker PAHM (2012) The rhizosphere microbiome and plant health. Trends Plant Sci 17(8):478–486. https://doi.org/10.1016/j.tplants.2012.04.001

Berg G, Rybakova D, Fischer D, Cernava T, Vergès M-CC, Charles T, Chen X, Cocolin L, Eversole K, Corral GH, Kazou M, Kinkel L, Lange L, Lima N, Loy A, Macklin JA, Maguin E, Mauchline T, McClure R et al (2020a) Microbiome definition re-visited: old concepts and new challenges. Microbiome 8(1):103. https://doi.org/10.1186/s40168-020-00875-0

Berg S, Dennis PG, Paungfoo-Lonhienne C, Anderson J, Robinson N, Brackin R, Royle A, DiBella L, Schmidt S (2020b) Effects of commercial microbial biostimulants on soil and root microbial communities and sugarcane yield. Biol Fertil Soils 56(4):565–580. https://doi.org/10.1007/s00374-019-01412-4

Bergez J-E, Audouin E, Therond O (n.d.) Agroecological transitions: from theory to practice in local participatory design

Blanco AJV, Costa MO, Silva R, Albuquerque FS, Melo ATO, Lopes FAC, Steindorff AS, Barbosa ET, Ulhoa CJ, Lobo Junior M (2018) Diversity and pathogenicity of Rhizoctonia species from the Brazilian Cerrado. Plant Dis 102(4):773–781. https://doi.org/10.1094/PDIS-05-17-0721-RE

Bonanomi G, Gaglione SA, Cesarano G, Sarker TC, Pascale M, Scala F, Zoina A (2017) Frequent applications of organic matter to agricultural soil increase fungistasis. Pedosphere 27(1):86–95. https://doi.org/10.1016/S1002-0160(17)60298-4

Bouchez T, Blieux AL, Dequiedt S, Domaizon I, Dufresne A, Ferreira S, Godon JJ, Hellal J, Joulian C, Quaiser A, Martin-Laurent F, Mauffret A, Monier JM, Peyret P, Schmitt-Koplin P, Sibourg O, D'oiron E, Bispo A, Deportes I et al (2016) Molecular microbiology methods for environmental diagnosis. Environ Chem Lett 14(4):423–441. https://doi.org/10.1007/s10311-016-0581-3

Bulgari R, Cocetta G, Trivellini A, Vernieri P, Ferrante A (2015) Biostimulants and crop responses: a review. Biol Agric Hortic 31(1):1–17. https://doi.org/10.1080/01448765.2014.964649

Bünemann EK, Bongiorno G, Bai Z, Creamer RE, De Deyn G, de Goede R, Fleskens L, Geissen V, Kuyper TW, Mäder P, Pulleman M, Sukkel W, van Groenigen JW, Brussaard L (2018) Soil quality – a critical review. Soil Biol Biochem 120:105–125. https://doi.org/10.1016/j.soilbio.2018.01.030

Calvo P, Nelson L, Kloepper JW (2014) Agricultural uses of plant biostimulants. Plant Soil 383(1–2):3–41. https://doi.org/10.1007/s11104-014-2131-8

Campbell CL (1994) Epidemiology and management of root diseases. Springer, Berlin Heidelberg

Castro Lopes AA, Sousa DMG, dos Reis FB, Figueiredo CC, Malaquias JV, Souza LM, Carvalho Mendes I (2018) Temporal variation and critical limits of microbial indicators in oxisols in the Cerrado, Brazil. Geoderma Reg 12:72–82. https://doi.org/10.1016/j.geodrs.2018.01.003

Chabrillat S, Ben-Dor E, Cierniewski J, Gomez C, Schmid T, van Wesemael B (2019) Imaging spectroscopy for soil mapping and monitoring. Surv Geophys 40(3):361–399. https://doi.org/10.1007/s10712-019-09524-0

Chagas LFB, Martins ALL, de Carvalho Filho MR, de Oliveira Miller L, de Oliveira JC, Chagas Junior AF (2018) Bacillus subtilis E Trichoderma sp. no incremento da biomassa em plantas de soja, feijão-caupi, milho e arroz. Agri-Environ Sci 3(2):10–18. https://doi.org/10.36725/agries.v3i2.430

Chang C-W, Laird DA, Mausbach MJ, Hurburgh CR (2001) Near-infrared reflectance spectroscopy–principal components regression analyses of soil properties. Soil Sci Soc Am J 65(2):480–490. https://doi.org/10.2136/sssaj2001.652480x

Chiaramonte JB, Mendes LW, Mendes R (2021) Rhizosphere microbiome and soil-borne diseases, pp 155–168. https://doi.org/10.1007/978-981-15-6125-2_7

Civardi EA, Görgen CA, Ragagnin VA, da Silveira Neto AN, Carneiro LC, Lobo M (2019) Management of Congo grass cover crop affects timing of Sclerotinia sclerotiorum carpogenic germination and decay of soybean stem rot. Tropic Plant Pathol 44(1):94–103. https://doi.org/10.1007/s40858-018-0268-3

Colla G, Hoagland L, Ruzzi M, Cardarelli M, Bonini P, Canaguier R, Rouphael Y (2017) Biostimulant action of protein hydrolysates: unraveling their effects on plant physiology and microbiome. Front Plant Sci 8. https://doi.org/10.3389/fpls.2017.02202

CONAB – Companhia Nacional de Abastecimento. Produção agrícola–safra: série histórica dos grãos (2020)

CONAB – Companhia Nacional de Abastecimento. Produção agrícola (2023)

CONAB – Companhia Nacional de Abastecimento. Produção agrícola Acompanhamento da safra Brasileira de Grãos (2024). https://www.conab.gov.br/info-agro/safras/graos/boletim-da-safra-de-graos

Da Cunha Leme Filho JF, Thomason WE, Evanylo GK, Zhang X, Strickland MS, Chim BK, Diatta AA (2021) An integrated plant nutrition system (IPNS) for corn in the mid-Atlantic USA. J Plant Nutr 44(5):704–722. https://doi.org/10.1080/01904167.2020.1849298

de Almeida TA, Giancotti PRF, de Lima BAD, Dalla Nora D, Gomes RR (2018) Biofertilizer microgeo® on rice crop: yield and seed quality. J Agric Sci 10(5):288. https://doi.org/10.5539/jas.v10n5p288

de Azevedo Silva F, de Oliveira Vieira V, Carrenho R, Rodrigues VB, Junior ML, da Silva GF, Soares MA (2021) Influence of the biocontrol agents Trichoderma spp. on the structure and functionality of the edaphic microbial community in common bean cultivars (Phaseolus vulgaris L.) inoculated with Sclerotinia sclerotiorum (Lib.) de Bary. Appl Soil Ecol 168:104190. https://doi.org/10.1016/j.apsoil.2021.104190

De Corato U (2020) Disease-suppressive compost enhances natural soil suppressiveness against soil-borne plant pathogens: a critical review. Rhizosphere 13:100192. https://doi.org/10.1016/j.rhisph.2020.100192

de Freitas PL, Landers JN (2014) The transformation of agriculture in Brazil through development and adoption of zero tillage conservation agriculture. Int Soil Water Conserv Res 2(1):35–46. https://doi.org/10.1016/S2095-6339(15)30012-5

de Souza R, Ambrosini A, Passaglia LMP (2015) Plant growth-promoting bacteria as inoculants in agricultural soils. Genet Mol Biol 38(4):401–419. https://doi.org/10.1590/S1415-475738420150053

de Toledo-Souza ED, da Silveira PM, Lobo Junior M, Café Filho AC (2008) Sistemas de cultivo, sucessões de culturas, densidade do solo e sobrevivência de patógenos de solo. Pesq Agrop Brasileira 43(8):971–978. https://doi.org/10.1590/S0100-204X2008000800004

de Toledo-Souza ED, da Silveira PM, Café-Filho AC, Lobo Junior M (2012) Fusarium wilt incidence and common bean yield according to the preceding crop and the soil tillage system. Pesq Agrop Brasileira 47(8):1031–1037. https://doi.org/10.1590/S0100-204X2012000800002

Demattê JAM, Paiva AFDS, Poppiel RR, Rosin NA, Ruiz LFC, Mello FADO, Minasny B, Grunwald S, Ge Y, Ben Dor E, Gholizadeh A, Gomez C, Chabrillat S, Francos N, Ayoubi S, Fiantis D, Biney JKM, Wang C, Belal A et al (2022) The Brazilian soil spectral service (BraSpecS): a user-friendly system for global soil spectra communication. Remote Sens 14(3):740. https://doi.org/10.3390/rs14030740

Doran JW, Sarrantonio M, Liebig MA (1996) Soil health and sustainability, pp 1–54. https://doi.org/10.1016/S0065-2113(08)60178-9

du Jardin P (2015) Plant biostimulants: definition, concept, main categories and regulation. Sci Hortic 196:3–14. https://doi.org/10.1016/j.scienta.2015.09.021

Eilenberg J, Hajek A, Lomer C (2001) Suggestions for unifying the terminology in biological control. BioControl 46(4):387–400. https://doi.org/10.1023/A:1014193329979

Evangelista Da Silva E, Sabadin De Azevedo PH, De-Polli H (n.d.) Determinação da respiração basal (RBS) e quociente metabólico do solo (qCO 2)

FAO (2009) Glossary on Organic Agriculture

FAO (2021) Program of Brazil-FAO Int'l Cooperation. https://www.fao.org

Foo JL, Ling H, Lee YS, Chang MW (2017) Microbiome engineering: current applications and its future. Biotechnol J 12(3):1600099. https://doi.org/10.1002/biot.201600099

Freitas G, Michelliny V, De Matos P, Yeda B, Malheiros De Oliveira M, Karla D, Marques S, Corrêa J, Silva B (n.d.) Sustainable development goal 15 LIFE ON LAND CONTRIBUTIONS OF EMBRAPA

Gandariasbeitia M, Besga G, Albizu I, Larregla S, Mendarte S (2017) Prediction of chemical and biological variables of soil in grazing areas with visible- and near-infrared spectroscopy. Geoderma 305:228–235. https://doi.org/10.1016/j.geoderma.2017.05.045

Garbeva P, van Veen JA, van Elsas JD (2004) Microbial diversity in soil: selection of microbial populations by plant and soil type and implications for disease suppressiveness. Annu Rev Phytopathol 42(1):243–270. https://doi.org/10.1146/annurev.phyto.42.012604.135455

Geraldine AM, Lopes FAC, Carvalho DDC, Barbosa ET, Rodrigues AR, Brandão RS, Ulhoa CJ, Lobo Junior M (2013) Cell wall-degrading enzymes and parasitism of sclerotia are key factors on field biocontrol of white mold by Trichoderma spp. Biol Control 67(3):308–316. https://doi.org/10.1016/j.biocontrol.2013.09.013

Ghini R, Bettiol W, Hamada E (2011) Diseases in tropical and plantation crops as affected by climate changes: current knowledge and perspectives. Plant Pathol 60(1):122–132. https://doi.org/10.1111/j.1365-3059.2010.02403.x

Glare TR, Nollet LML (2023) Types of biopesticides. In: Biopesticides handbook. CRC Press, pp 7–24. https://doi.org/10.1201/9781003265139-3

Goulet F (2021) Characterizing alignments in socio-technical transitions. Lessons from agricultural bio-inputs in Brazil. Technol Soc 65:101580. https://doi.org/10.1016/j.techsoc.2021.101580

Hartmann M, Frey B, Mayer J, Mäder P, Widmer F (2015) Distinct soil microbial diversity under long-term organic and conventional farming. ISME J 9(5):1177–1194. https://doi.org/10.1038/ismej.2014.210

Hassan Dar G, Zargar MY, Beigh GM (1997) Biocontrol of Fusarium Root Rot in the common bean (Phaseolus vulgaris L.) by using symbiotic glomus mosseae and rhizobium leguminosarum. Microb Ecol 34(1):74–80. https://doi.org/10.1007/s002489900036

Hungria M, Franchini JC, Brandão-Junior O, Kaschuk G, Souza RA (2009) Soil microbial activity and crop sustainability in a long-term experiment with three soil-tillage and two crop-rotation systems. Appl Soil Ecol 42(3):288–296. https://doi.org/10.1016/j.apsoil.2009.05.005

Janvier C, Villeneuve F, Alabouvette C, Edel-Hermann V, Mateille T, Steinberg C (2007) Soil health through soil disease suppression: which strategy from descriptors to indicators? Soil Biol Biochem 39(1):1–23. https://doi.org/10.1016/j.soilbio.2006.07.001

Karlen DL, Mausbach MJ, Doran JW, Cline RG, Harris RF, Schuman GE (1997) Soil quality: a concept, definition, and framework for evaluation (a guest editorial). Soil Sci Soc Am J 61(1):4–10. https://doi.org/10.2136/sssaj1997.03615995006100010001x

Karlen DL, Veum KS, Sudduth KA, Obrycki JF, Nunes MR (2019) Soil health assessment: past accomplishments, current activities, and future opportunities. Soil Tillage Res 195:104365. https://doi.org/10.1016/j.still.2019.104365

Kassam A, Friedrich T, Derpsch R (2019) Global spread of conservation agriculture. Int J Environ Stud 76(1):29–51. https://doi.org/10.1080/00207233.2018.1494927

Ke J, Wang B, Yoshikuni Y (2021) Microbiome engineering: synthetic biology of plant-associated microbiomes in sustainable agriculture. Trends Biotechnol 39(3):244–261. https://doi.org/10.1016/j.tibtech.2020.07.008

Lehman R, Cambardella C, Stott D, Acosta-Martinez V, Manter D, Buyer J, Maul J, Smith J, Collins H, Halvorson J, Kremer R, Lundgren J, Ducey T, Jin V, Karlen D (2015) Understanding and enhancing soil biological health: the solution for reversing soil degradation. Sustainability 7(1):988–1027. https://doi.org/10.3390/su7010988

Lobo M, Macedo JR, Ferreira P, Santos-Goulart D (n.d.) Solos supressivos a doenças, 2022

Madden LV, Hughes G, van den Bosch F (2007) The study of plant disease epidemics. The American Phytopathological Society. https://doi.org/10.1094/9780890545058

MAPA – Ministério da Agricultura e Pecuária. Principais marcos históricos dos 160 anos do MAPA (2020) MAPA

Martins Alves N, Araújo Guimarães R, Silva Costa Guimarães S, Frausino de Faria A, Augusto Férrer Melo Santos I, Henrique Vasconcelos de Medeiros F, Jank L, Gomes Cardoso P (2021) A Trojan horse approach for white mold biocontrol: Paraconiothyrium endophytes promotes grass growth and inhibits Sclerotinia sclerotiorum. Biol Control 160:104685. https://doi.org/10.1016/j.biocontrol.2021.104685

McSpadden Gardener BB, Weller DM (2001) Changes in populations of rhizosphere bacteria associated with take-all disease of wheat. Appl Environ Microbiol 67(10):4414–4425. https://doi.org/10.1128/AEM.67.10.4414-4425.2001

Mendes R, Kruijt M, de Bruijn I, Dekkers E, van der Voort M, Schneider JHM, Piceno YM, DeSantis TZ, Andersen GL, Bakker PAHM, Raaijmakers JM (2011) Deciphering the rhizosphere microbiome for disease-suppressive bacteria. Science 332(6033):1097–1100. https://doi.org/10.1126/science.1203980

Mendes LW, Kuramae EE, Navarrete AA, van Veen JA, Tsai SM (2014) Taxonomical and functional microbial community selection in soybean rhizosphere. ISME J 8(8):1577–1587. https://doi.org/10.1038/ismej.2014.17

Mendes IC, Tormena CA, Cherubin MR, Karlen DL (2018) Soil health assessment and maintenance in Central and South-Central Brazil, pp 379–416. https://doi.org/10.19103/AS.2017.0033.35

Meyer MC, Campos HD, Godoy CV, Utiamada CM, Dias AR, De Souza D, Filho J, Lopes De Medeiros FC, Galdino JV, Nunes Junior J, Henrique L, Pereira Da Silva C, Sato N, Neves De Oliveira MC, Martins C, Tormen R (n.d.) Eficiência de fungicidas para controle de mofo-branco

(Sclerotinia sclerotiorum) em soja, na safra 2018/19: resultados sumarizados dos experimentos cooperativos

Michereff, S. J., Andrade, D. E. G. T., & Menezes, M. (2005). Ecologia e Manejo de Patógenos Radiculares em Solos Tropicais

Minaxi, Nain L, Yadav RC, Saxena J (2012) Characterization of multifaced *Bacillus* sp RM-2 for its use as plant growth-promoting bioincoculant for cropd grown in semi-arid deserts. Appl Soil Ecol 59:124–139

Moreira FMS, Siqueira JO (2006) Microbiologia e bioquímica do solo, 2nd edn. UFLA

Nain L, Yadav RC, Saxena J (2012) Characterization of multifaceted Bacillus sp. RM-2 for its use as plant growth promoting bioinoculant for crops grown in semi arid deserts. Appl Soil Ecol 59:124–135. https://doi.org/10.1016/j.apsoil.2011.08.001

Nascimento VC, Rodrigues-Santos KC, Carvalho-Alencar KL, Castro MB, Kruger RH, Lopes FAC (2022) Trichoderma: biological control efficiency and perspectives for the Brazilian Midwest states and Tocantins. Braz J Biol 82. https://doi.org/10.1590/1519-6984.260161

Nosheen S, Ajmal I, Song Y (2021) Microbes as biofertilizers, a potential approach for sustainable crop production. Sustainability 13(4):1868. https://doi.org/10.3390/su13041868

Nunes PSO, Lacerda-Junior GV, Mascarin GM, Guimarães RA, Medeiros FHV, Arthurs S, Bettiol W (2024) Microbial consortia of biological products: do they have a future? Biol Control 188:105439. https://doi.org/10.1016/j.biocontrol.2024.105439

Parker IM, Gilbert GS (2018) Density-dependent disease, life-history trade-offs, and the effect of leaf pathogens on a suite of co-occurring close relatives. J Ecol 106(5):1829–1838. https://doi.org/10.1111/1365-2745.13024

Pathogens, precipitation and produce prices (2021) Nature climate change. Nat Res 11(8):635. https://doi.org/10.1038/s41558-021-01124-4

Paustian K, Collier S, Baldock J, Burgess R, Creque J, DeLonge M, Dungait J, Ellert B, Frank S, Goddard T, Govaerts B, Grundy M, Henning M, Izaurralde RC, Madaras M, McConkey B, Porzig E, Rice C, Searle R et al (2019) Quantifying carbon for agricultural soil management: from the current status toward a global soil information system. Carbon Management 10(6):567–587. https://doi.org/10.1080/17583004.2019.1633231

Paz-Kagan T, Shachak M, Zaady E, Karnieli A (2014) A spectral soil quality index (SSQI) for characterizing soil function in areas of changed land use. Geoderma 230–231:171–184. https://doi.org/10.1016/j.geoderma.2014.04.003

Pietikäinen J, Fritze H (1995) Clear-cutting and prescribed burning in coniferous forest: comparison of effects on soil fungal and total microbial biomass, respiration activity and nitrification. Soil Biol Biochem 27(1):101–109. https://doi.org/10.1016/0038-0717(94)00125-K

Posso Suárez DF, Mattos ADP, Broti Rissato B, Schwan-Estrada KRF (2020) Activación de mecanismos de defensa en maíz pira mediante el uso del abono orgánico Microgeo®. Rev Mex Cienc Agríc 11(5):965–977. https://doi.org/10.29312/remexca.v11i5.2009

Postma J, Schilder MT, Bloem J, van Leeuwen-Haagsma WK (2008) Soil suppressiveness and functional diversity of the soil microflora in organic farming systems. Soil Biol Biochem 40(9):2394–2406. https://doi.org/10.1016/j.soilbio.2008.05.023

Poudel R, Jumpponen A, Schlatter DC, Paulitz TC, Gardener BBM, Kinkel LL, Garrett KA (2016) Microbiome networks: a systems framework for identifying candidate microbial assemblages for disease management. Phytopathology 106(10):1083–1096. https://doi.org/10.1094/PHYTO-02-16-0058-FI

Raaijmakers JM, Mazzola M (2016) Soil immune responses. Science 352(6292):1392–1393. https://doi.org/10.1126/science.aaf3252

Rangel da Silva Gama D, Nascimento Santos E, Santos Dourado Ferreira T, Carlos Mesquita A, Eishi Yuri J (2014) Utilização de biofertilizante na produção de melão do tipo amarelo

Raza W, Ling N, Zhang R, Huang Q, Xu Y, Shen Q (2017) Success evaluation of the biological control of *Fusarium* wilts of cucumber, banana, and tomato since 2000 and future research strategies. Crit Rev Biotechnol 37(2):202–212. https://doi.org/10.3109/07388551.2015.1130683

Rhodes CJ (2017) The imperative for regenerative agriculture. Science Progress 100(1):80–129. Science Reviews 2000 Ltd. https://doi.org/10.3184/003685017X14876775256165

Ricci M, Tilbury L, Daridon B, Sukalac K (2019) General principles to justify plant biostimulant claims. Front Plant Sci 10. https://doi.org/10.3389/fpls.2019.00494

Rodrigues FA, Corrêa GF, dos Santos MA, Borges Filho EL (1998) Fatores envolvidos na supressividade a Rhizoctonia solani em alguns solos tropicais brasileiros. Rev Bras Ciênc Solo 22(2):239–246. https://doi.org/10.1590/S0100-06831998000200009

Rouphael Y, Colla G (2020) Editorial: biostimulants in agriculture. Front Plant Sci 11:40. https://doi.org/10.3389/fpls.2020.00040

Silva AMM, de Araújo VLVP, Cardoso EJBN (2023) Revisiting the past to understand the present and future of soil health in Brazil. Front Soil Sci 3. https://doi.org/10.3389/fsoil.2023.1172436

Singh BK, Trivedi P, Egidi E, Macdonald CA, Delgado-Baquerizo M (2020) Crop microbiome and sustainable agriculture. Nat Rev Microbiol 18(11):601–602. https://doi.org/10.1038/s41579-020-00446-y

Sobucki L, Ramos RF, Meireles LA, Antoniolli ZI, Jacques RJS (2021) Contribution of enzymes to soil quality and the evolution of research in Brazil. Rev Bras Ciênc Solo 45. https://doi.org/10.36783/18069657rbcs20210109

Srivastava AK, Arora DK, Gupta S, Pandey RR, Lee MW (1996) Diversity of potential microbial parasites colonizing sclerotia of Macrophomina phaseolina in soil. Biol Fertil Soils 22(1–2):136–140. https://doi.org/10.1007/BF00384445

Suarez A, Molina RS, Ramponi G, Petrina R, Bollatti L, Sequeiros D (2020) Pest detection and classification to reduce pesticide use in fruit crops based on deep neural networks and image processing. Arts units.it.

Topalović O, Hussain M, Heuer H (2020) Plants and associated soil microbiota cooperatively suppress plant-parasitic nematodes. Front Microbiol 11. https://doi.org/10.3389/fmicb.2020.00313

Velásquez AC, Castroverde CDM, He SY (2018) Plant–pathogen warfare under changing climate conditions. Curr Biol 28(10):R619–R634. https://doi.org/10.1016/j.cub.2018.03.054

Vessey JK (2003) Plant growth promoting rhizobacteria as biofertilizers. Plant Soil 255(2):571–586. https://doi.org/10.1023/A:1026037216893

Veum KS, Sudduth KA, Kremer RJ, Kitchen NR (2015) Estimating a soil quality index with VNIR reflectance spectroscopy. Soil Sci Soc Am J 79(2):637–649. https://doi.org/10.2136/sssaj2014.09.0390

Watson AG, Ford EJ (1972) Soil Fungistasis – a reappraisal. Annu Rev Phytopathol 10(1):327–346. https://doi.org/10.1146/annurev.py.10.090172.001551

Wen L, Lee-Marzano S, Ortiz-Ribbing LM, Gruver J, Hartman GL, Eastburn DM (2017) Suppression of soilborne diseases of soybean with cover crops. Plant Dis 101(11):1918–1928. https://doi.org/10.1094/PDIS-07-16-1067-RE

Westphal A (2005) Detection and description of soils with specific nematode suppressiveness. J Nematol 37(1):121–132

Yakhin OI, Lubyanov AA, Yakhin IA, Brown PH (2017) Biostimulants in plant science: a global perspective. Front Plant Sci 7:2049. https://doi.org/10.3389/fpls.2016.02049

Yang H, Li J, Xiao Y, Gu Y, Liu H, Liang Y, Liu X, Hu J, Meng D, Yin H (2017) An integrated insight into the relationship between soil microbial community and tobacco bacterial wilt disease. Front Microbiol 8. https://doi.org/10.3389/fmicb.2017.02179

Chapter 10
Microbiomes in Suppressive Soils and Its Implications on Plant Pathogens

L. Rajendran, K. Saravanakumari, Jothini Varsha, V. Sendhilvel, and Somasundaram Jayaraman

1 Introduction

Soil is a fundamental material covering much of the land surface for farming, and is made up of inorganic particles and organic matter. It provides structural support, water, and nutrients essential for plants, especially in agriculture. Soil properties can vary widely due to factors such as leaching, weathering, and microbial activity, resulting in different soil types with specific advantages and limitations for farming. Soil is defined as a natural entity comprising solids (minerals and organic matter), liquids, and gases that exists on the land surface, occupies space, and is characterized by distinct horizons or layers formed through the addition, loss, transfer, and transformation of energy and matter (Brady 1996). It must also be capable of supporting rooted plants in a natural setting. The upper boundary of soil is where it meets air, shallow water, live plants, or undecomposed plant materials. Areas where the surface is permanently submerged under water deeper than about 2.5 meters are not considered soil.

The lower boundary of soil, distinguishing it from the underlying non-soil material, is more challenging to define. Soil typically transitions to hard rock or to materials lacking biological activity. For classification purposes, this lower boundary is conventionally set at 200 cm below the surface. Disease suppressive soils have been defined by Baker and Cook (1974) as "soils in which the pathogen does not establish or persist, establishes but causes little or no damage, or establishes and causes disease for a while but thereafter the disease is less important, although the pathogen may persist in the soils". In contrast, diseases readily occur in conducive soils

L. Rajendran · K. Saravanakumari · J. Varsha · V. Sendhilvel
Department of Plant Pathology, Tamil Nadu Agricultural University, Coimbatore, India

S. Jayaraman (✉)
ICAR-Indian Institute of Soil and Water Conservation Research Centre, Fernhill, Udhagamandalam, Tamil Nadu, India

S. Gnanamanickam, M. Höfte (eds.), *Disease Suppressive Tropical Soils*,
https://doi.org/10.1007/978-3-032-08100-1_10

in which abiotic and biotic conditions are favourable to pathogens. Thus, a clearer understanding of the disease inhibitory mechanisms of suppressive soils will lead to the development of powerful tools to control diseases. Pathogens like *Fusarium oxysporum*, *Gaeumannomyces graminis* var. *tritici*, and plant parasitic nematodes such as *Heterodera*, *Globodera*, *Meloidogyne*, along with *Rhizoctonia solani*, *Thielaviopsis basicola*, and *Verticillium dahliae*, have garnered significant attention. Particularly noteworthy are soils that naturally suppress to *G. graminis* var. *tritici*, *F. oxysporum*, and *R. solani*, areas that have been extensively researched. The root disease take-all, caused by *Gaeumannomyces graminis* var. *tritici*, is globally recognized as the most economically significant threat to wheat. Natural reductions in the occurrence and intensity of take-all have been observed following severe outbreaks during prolonged monocultures of wheat or barley. Suppressive soils offer prime examples of natural, microbe-based plant defense mechanisms where plant roots engage in rhizodeposition releasing substances that stimulate, enrich, and support soil microorganisms to act as the first line of defense against soil-borne pathogens (Weller et al. 2002, 2007). Such soils are characterized by their ability to prevent pathogen establishment or persistence, or to cause an initial disease outbreak that diminishes with successive plantings of a susceptible host, even if the pathogens remain in the soil. This suppressive activity is attributed to a combination of "general" and "specific" suppression mechanisms. General suppression refers to the soil's overall capacity to limit pathogen growth, its virulence and activity through the collective competitive and antagonistic effects of the entire soil microbiome and its composition against the plant pathogens (Cook 2014a, b). Here, the microbiome is defined as a community of diverse microorganisms present in a specific environment, such as the soil and rhizosphere, encompassing not just their genes, but also their transcripts, metabolites, and proteins. In case if specific suppression, microorganisms often target pathogens directly or occupy overlapping niches with them. For instance, *Dactylella oviparasitica* acts directly against the beet cyst nematode *Heterodera schachtii* (Olatinwo et al. 2006). Similarly, certain microorganisms demonstrate intimate niche overlap with pathogens, particularly in infection sites. Example includes *Agrobacterium radiobacter* competing with *Agrobacterium tumefaciens*, *Pseudomonas brassicacearum* interacting with *Gaeumannomyces graminis* var. *tritici*, and the interplay between pathogenic and non-pathogenic strains of *Fusarium oxysporum* (Weller et al. 2007).

In this chapter, we will examine three contemporary model systems that exemplify features of both 'specific' and 'general' disease suppression. Using these examples, we will explore hypotheses regarding the fundamental characteristics of microbial communities in the disease-suppressive soils, analyze and evaluate how new techniques offer innovative methods for testing these hypotheses, and ultimately, discuss how these insights can advance phytobiome-based approaches for sustainable disease management in agriculture. Our aim is to demonstrate how understanding indigenous soil microbes can be harnessed to enhance plant pathogen suppression and support sustainable agricultural practices.

2 Soils and Microbes: Intricacies and Relationship

Soil is a multifaceted and largest ecosystem that provides nourishment and provides housing to billions of microorganisms. In a single teaspoon of soil, the number of microbes vastly exceeds the human population on Earth. Soils host approximately 8 to 15 tons of microorganisms (at 0–15 cm depth), including bacteria, fungi, protozoa, nematodes, earthworms, and arthropods. Details on the quantity and biomass of these microbial species are provided in Table 10.1 (Hoorman 2020).

It has a diverse range of microbiome (Guerra et al. 2020). The range of elements present in solid fractions including sand, silt, clay, and other organic matter provides distinct microhabitats (Van Elsas and Trevors 1997). These microorganisms combinedly act as agents for various soil processes and one such processes that includes disease suppression (Garbeva et al. 2004). It is a natural and general action present in many soils and has the ability to limit the pathogen growth, virulence, inoculums and activity of the disease causing pathogens. This is achieved through the collective competitive and antagonistic mechanism of the entire soil microbiome against plant pathogens (Cook 2014a, b).

Suppressive soil has been categorized into two categories,

I. **General soil suppression**: a larger microbiome which inhibits broad range of plant pathogens (Cook and Baker 1983) Eg: Control of *Pythium, Fusarium* and *Phytophthora* by application of organic amendments (Weller et al. 2002).
II. **Specific soil suppression**: a selective group of microbes inhibits only particular pathogen (Weller et al. 2002). Eg: Take all disease of wheat caused by *Gaeumannomyces graminis* var. *tritici*, suppressed by *Pseudomonas* spp. by accumulation of 2,4-diacetylphloroglucinol (2,4-DAPG) (Kwak and Weller 2013).

3 Microbial Communities

Soil biota includes a vast array of organisms, among which microorganisms such as bacteria, fungi, and archaea are the most numerous and biomass-dominant. These soil microbes are crucial for various ecosystem functions and play a key role in

Table 10.1 Relative number and biomass of microbial species at 0–15 cm depth of soil

Microorganisms	Number/g of soil	Biomass (g/m^2)
Fungi	10^5–10^6	100–1500
Bacteria	10^8–10^9	40–500
Algae	10^4–10^5	1–50
Nematodes	10^2–10^3	Varies
Protozoa	10^3–10^4	Varies
Actinomycetes	10^7–10^8	40–500

regulating important processes such as carbon and nitrogen cycles. For instance, they are responsible for vital functions such as N_2-fixation, ammonia oxidation, denitrification, and ammonification. Microbial communities also contribute to the breakdown of different compounds, including organic pollutants like pesticides, and support plant growth and disease management (Wu et al. 2008). The diversity and abundance of soil microbes are critical for maintaining fundamental ecosystem processes and supporting crop production. Healthy soil, characterized by a diverse and active microbial community, is essential for plant growth and, consequently, for successful crop yields. Soil microbial biomass, activity, and diversity serve as indicators of soil fertility and overall ecosystem productivity, and are therefore used to assess soil quality and health.

Soil quality is defined as the ability of a soil to function effectively within its natural or managed ecosystem to support plant and animal productivity, enhance water and air quality, and promote human health and habitation. Microbial populations in the soil can vary based on factors such as soil type, plant presence, and climate, leading to different responses to fertilization treatments depending on these abiotic conditions. Over time, soil, plants, and microbes have evolved together to form stable interactions within ecosystems, leading to the concept of the soil microbiome (Dinca et al. 2022). The extensive variety of living microorganisms in soil can be collectively referred to as "soil microbial biomass." This concept has been valuable for modeling the cycling of nutrients and energy. Microbially-mediated activities in soil encompass both beneficial aspects, such as carbon and nitrogen cycling and symbiotic relationships with plants, as well as harmful effects, like nitrous oxide (N_2O) and methane (CH_4) emissions and plant diseases. Understanding these activities is crucial for effective soil management and reducing negative environmental impacts. To achieve this, we need a more detailed and predictive understanding of the microbial communities involved and their responses to environmental stress and climate change. However, studying soil microorganisms is challenging due to the variability in soil composition, the enormous quantities (approximately 10^9 individual cells) and diversity (over 10^6 distinct taxa) found in each gram of soil, and the limited knowledge about most of these microbial populations. In fact, only about 1% of soil bacteria can typically be cultured in laboratory media.

Microbial community composition of a specific disease-suppressive soil differed with its disease-conducive soil, which was documented for several soil-borne diseases including potato scab, Fusarium wilt of banana, apple replant disease, wheat take-all and tobacco black root rot. Suppressive soil will minimize the risk of total crop loss and pesticide requirements. Only about 1% of the soil bacterial cells can be cultured using standard laboratory culturing techniques and the remaining 99% of them are unculturable. In the era of biotechnology, genomes of unculturable microbes from environmental samples are being estimated by culture-independent molecular techniques. Ng et al. (2011) utilized the metagenomic approach to describe the DNA of viral community from whiteflies (*Bemisia tabaci*). Molecular techniques using stable markers like 16S rRNA provide relatively new avenues for studying soil microbial communities. The community profling of an ecosystem can be resolved from metagenomic DNA through sequencing of 16S rRNA gene clonal libraries. The sequence-driven approach has been successfully applied to distinguish the bacterial diversity between disease-suppressive and disease-conducive soils (Kyselkova et al. 2009). Assessment of microbial diversity in disease-suppressive soil is essential to recognize the contribution of dominant beneficial microbial groups that cause disease suppressiveness (Borneman and Becker 2007). Mendes et al. (2011) analysed the disease-suppressive soil by culture-dependent functional analyses and found that upon attack by a fungal root pathogen, plants can exploit microbial consortia from soil for protection against infections. Similarly, Penton et al. (2014) analysed Rhizoctonia-suppressive soil for fungal community structure using 28S LSU gene sequencing and identified that the majority of differences associated with suppressive soils were attributed to a number of endophytic species with plant pathogen suppression potentials and mycoparasites, whereas non-suppressive soils were dominated by plant pathogen communities

A culture-independent molecular technique was employed to analyze the bacterial communities in suppressive and conducive soils. Metagenomic DNA was extracted directly from both soil types, and the sequence polymorphism was assessed by targeting the hypervariable V4 and V5 regions of the 16S rRNA gene. The findings from the 16S rRNA gene analysis, combined with the assessment of soil physico-chemical and biological properties, effectively differentiated *S. rolfsii* suppressive soil from conducive soil. In suppressive soil, Actinobacteria was the dominant phylogenetic group, followed by Proteobacteria, with additional presence of Acidobacteria, Firmicutes, and Cyanobacteria. In contrast, conducive soil exhibited a low abundance of Actinobacteria and was primarily dominated by Gamma- and Betaproteobacteria. Overall, the suppressive soil showed greater bacterial diversity and species richness compared to the conducive soil. This study identifies the key bacterial communities associated with the suppression of *S. rolfsii* (Thilagavathi et al. 2021). Fluorescent pseudomonads isolated from a soil that suppresses Fusarium wilt were found to significantly reduce take-all disease (caused by *Gaeumannomyces graminis* var. *tritici*) in wheat and *Ophiobolus* patch (caused by *G. graminis* var. *avenae*) in Agrostis turfgrass. The bacteria were applied to the soil at a concentration of 10^7 colony-forming units (cfu) per gram at the time of sowing. Results showed a significant reduction ($P \leq 0.05$) in diseased wheat roots when both the bacteria and pathogen were present, compared to the pathogen alone. Additionally, the dry weights of the tops of wheat and Agrostis turfgrass were significantly higher ($P \leq 0.01$) in treatments with the bacteria and pathogens compared to controls with pathogens alone. However, the dry weights of plants treated with the bacteria alone were not significantly different from those of healthy, non-inoculated wheat plants, indicating that the fluorescent pseudomonads did not promote plant growth. At the conclusion of the experiments the bacterial isolates, which were genetically marked with rifampicin resistance, were recovered from the wheat roots and rhizosphere soil at concentrations ranging from 10^5 to 10^7 cfu per gram of fresh root weight or oven-dried rhizosphere soil (Wong and Baker 1984).

The microbial communities involved in disease suppression were influenced by the following factors:

- Host plant and cultivar: provides carbon and other energy supplements
- Soil type: key nutrients availability based on texture, structure and other parameters like pH
- Agronomic practices: soil cultivation, cropping pattern, alternating crops, use of herbicides and fertilizers etc. (Garbeva et al. 2004).

4 Soil Borne Pathogens and Beneficial Microbes

Soil suppressiveness refers to the inherent ability of soil to inhibit soil-borne pathogens and promote root growth. For over 70 years, it has been recognized that soils with natural disease-suppressive properties are primarily influenced by beneficial microorganisms. These microorganisms interact with plant pathogens, impacting their survival, growth, and ability to infect plants. There are two main types of soil suppressiveness: general suppressiveness, which is based on the overall microbial community, and specific suppressiveness, which is linked to particular microbial populations. General suppressiveness relies on the entire soil microbial biomass, which inhibits pathogens through various mechanisms, particularly competition and the release of microbial inhibitors. This type of suppressiveness cannot be experimentally transferred between soils. Consequently, most soils exhibit some degree of general suppressiveness against soil-borne diseases, with the extent influenced by soil type, farming practices, and overall microbial activity. General suppressiveness often prevents pathogens from surviving and proliferating in the soil, a phenomenon known as fungistasis when dealing with fungal pathogens.

Fungistasis can impact *Fusarium* pathogens, though its role in relation to different *Fusarium* species or formae speciales requires further investigation. Research conducted on 31 wheat fields showed that soils exhibiting fungistasis against *F. graminearum* had higher bacterial diversity, more prevalent *Pseudomonas* and *Bacillus* species, and a more interconnected bacterial community. This suggests that interactions within diverse bacterial communities, including antagonistic taxa, play a crucial role in controlling *F. graminearum*. Similarly, another study by found differences in bacterial and fungal communities between Fusarium wilt-diseased soils and healthy (presumably suppressive) soils from various countries and crops. In addition to general suppressiveness, specific suppressiveness targets certain diseases and depends on the activity of specific plant-protecting microbial groups. Non-suppressive soils (also known as conducive soils) can be transformed into suppressive soils by inoculating them with a small percentage (0.1–10%) of suppressive soil. Although abiotic factors like soil physico-chemical properties may influence pathogen control, specific suppressiveness is primarily driven by beneficial soil microorganisms, as sterilization can turn suppressive soils into conducive ones.

Specific suppressiveness typically involves a few key microbial groups that protect plants, though comparisons of microbial communities between suppressive and conducive soils often reveal significant differences across a wide range of taxa.

Disease suppressiveness has been documented for various soil-borne fungal pathogens, including *Gaeumannomyces graminis* var. *tritici*, *Thievalopsis basicola*, and *Rhizoctonia solani*. It is particularly well-established for several Fusarium species, such as *F. culmorum* on wheat in the Netherlands and Germany, barley in Denmark, and cape gooseberry in Colombia, as well as *F. oxysporum* f. sp. *fragariae* on strawberry in Korea. This indicates that suppressiveness is observed across a diverse array of *Fusarium* pathosystems and is found in many regions around the world.

5 Beneficial Microbes

Plants benefit from various microbial populations inhabiting different environments around them. The rhizosphere- the soil region surrounding plant roots, is particularly favored by microbial communities. Some microbes are found on the plant surface (phyllosphere) as epiphytes, or inside the plant as endophytes (Weller et al. 2002). A key factor for microbial presence in soil is the organic matter loss from plant roots, which provides energy for microbial growth in the rhizosphere. This carbon loss attracts numerous microbial species, including saprotrophs, biotrophs like mycorrhizal fungi, and potentially harmful pathogens that can cause diseases. While microbes are present in various plant-associated environments, the rhizosphere is the most extensively studied due to its rich microbial diversity. Recent research has highlighted the different environmental niches occupied by microbes around the rhizosphere and their potential biotechnological applications. Microbial communities play a crucial role in addressing agroecological issues, enhancing disease resistance, managing stress, and reducing economic losses, thereby boosting agricultural productivity on a global scale.

Among plant-associated microorganisms, Plant Growth-Promoting Rhizobacteria (PGPR) are particularly notable. These bacteria, residing in the rhizosphere, promote plant health and growth through various mechanisms. Common PGPR genera include *Acinetobacter, Agrobacterium, Arthrobacter, Azotobacter, Azospirillum, Burkholderia, Bradyrhizobium, Rhizobium, Frankia, Serratia, Thiobacillus, Pseudomonas*, and *Bacillus*. Microbial symbionts can provide stress tolerance to a wide range of plants, both dicots and monocots. For example, *Pseudomonas* and *Burkholderia* species isolated from roots have demonstrated growth promotion and biocontrol activities. Additionally, Streptomyces mutabilis strains IA1 from Saharan soil have shown growth-promoting effects on wheat seedlings. Plant diseases are caused by a range of pathogens including fungi, bacteria, viruses, and nematodes. These diseases can be managed using biocontrol agents, which employ various mechanisms to suppress pathogens. These methods include directly parasitizing the pathogens, inhibiting them through the production of antibiotics and extracellular enzymes that degrade cell walls, competing for essential nutrients such as iron, nitrogen, or carbon at colonization sites, and stimulating the plant's own defense mechanisms to enhance resistance (Table 10.2).

Table 10.2 Plant Growth Promoting Microbes and their role

Plant growth promoting microbes	Origin	Properties
Agrobacterium radiobacter	Soil near peach tree	PGP/ Biocontrol
Azospirillum brasilense	Soil under soy filed	N-fixation
Bacillus amyloliquefacines	Plant pathogen infested soil	PGP/ Biocontrol
Bacillus pumilus	Soil	PGP
Bacillus subtilis	Soil	PGP/Biocontrol
Pseudomonas granadensis	Soil	PGP
Trichoderma harzianum	Soil	PGP/Biocontrol
Chaetomium globosum	Soil	PGP/Biocontrol

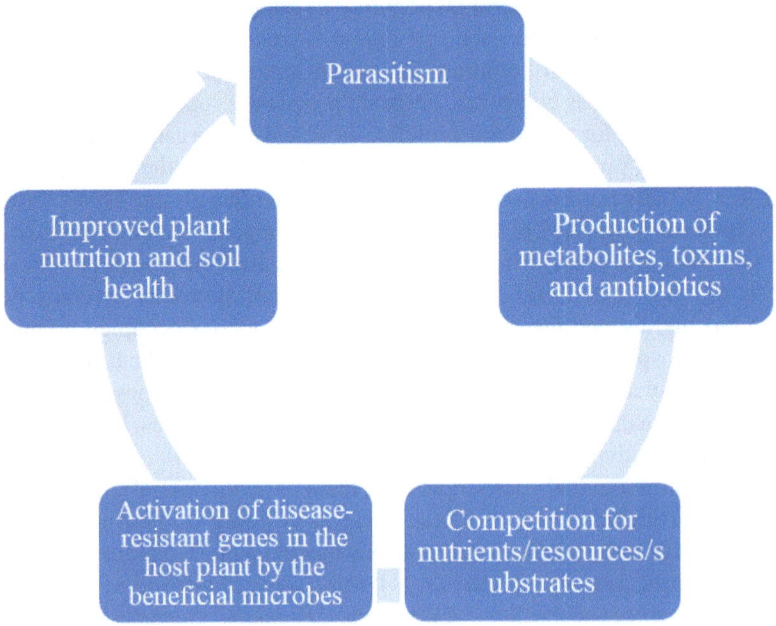

Fig. 10.1 Mechanism involved in suppression of soil borne diseases (Jayaraman et al. 2021)

Bacterial and Archaeal type microbes: Mendes et al. (2011) found that Proteobacteria, Firmicutes, and Actinobacteria exhibit more disease-suppressive traits. *Streptomyces, Bradyrhizobium, Burkholderia* (Brader et al. 2014; Kyselková and Moënne-Loccoz 2012), *Xylariaceae, Lactobacillaceae, Bacillus* (Penton et al. 2014; Wu et al. 2008), *Enterobacter* spp., *Flavobacterium aluminium, Pseudomonas* spp., *Streptomyces* spp. Fungal type microbes: *Penicillium* spp., *Trichoderma* spp., (Hoitink et al. 1997) Arbuscular mycorrhizal (AM) fungi (Li et al. 2007), non pathogenic *Alternaria, Gibberella* (Penton et al. 2014), *Aspergillus fumigatus* and *Fusarium solani* (Fig. 10.1).

Bacterial type microbes	Reference	Fungal type microbes	Reference
Streptomyces, *Bradyrhizobium* and *Burkholderia*	Brader et al. (2014), Kyselková and Moënne-Loccoz (2012)	*Penicillium* spp. and *Trichoderma* spp.	Hoitink et al. (1997)
Xylariaceae, Lactobacillaceae and *Bacillus*	Penton et al. (2014), Wu et al. (2008)	Arbuscular mycorrhizal (AM) fungi	Li et al. (2007)
Enterobacter spp., Flavobacterium aluminium, Pseudomonas spp. and *Streptomyces spp.*	Jayaraman et al. (2021)	Non pathogenic *Alternaria* and *Gibberella*	Penton et al. (2014)
Pseudomonas lurida FGD5-2, P. koreensis HCH2-3, and P. rhodesiae	Zheng et al. (2021)	*Aspergillus fumigatus* and *Fusarium solani*	Yuan et al. (2021)

6 Soil Health and Tritrophic Interactions

The rhizosphere environment is an important area for interaction between the plants (root exudates), soil, and microbes. These interactions between the microbial community in soil play a significant role in preventing the growth of pathogens and enhancing the plant's ability to resist diseases (Duran et al. 2018). The concept of disease suppression is visualized with a disease triangle with three determinants; host plant, soil-borne pathogen, and environmental factors (Scholthof 2007). Diseases occur when all the determinants are in favourable with the pathogen. Thus, the environment component is manipulated to develop disease suppressive soil or reduce the activity of disease development even in the presence of pathogen and host (Hadar and Papadopoulou 2012). Therefore, it is important to maintain the soil health as the disease suppression is linked to soil physical conditions, fertility level, biodiversity, populations of soil micro-organisms, and its soil management practices (Fig. 10.2).

7 Microbiome in Suppressive Soils and Metagenomics

Since the 1970s, research into soil microbiota and its role in suppressing soil-borne pathogens has advanced significantly. Studies have shown that certain soils possess natural suppressive abilities that can control plant pathogens, leading to increased interest in identifying various types of disease-suppressive soils. Recent research has highlighted that the microbiota in specific soils can create unfavorable conditions for pathogen development, thus enhancing the soil's suppressive capacity. This suppression is largely driven by the soil's microbial community and its organic matter, which also contribute to better plant nutrition and overall vigor. Many

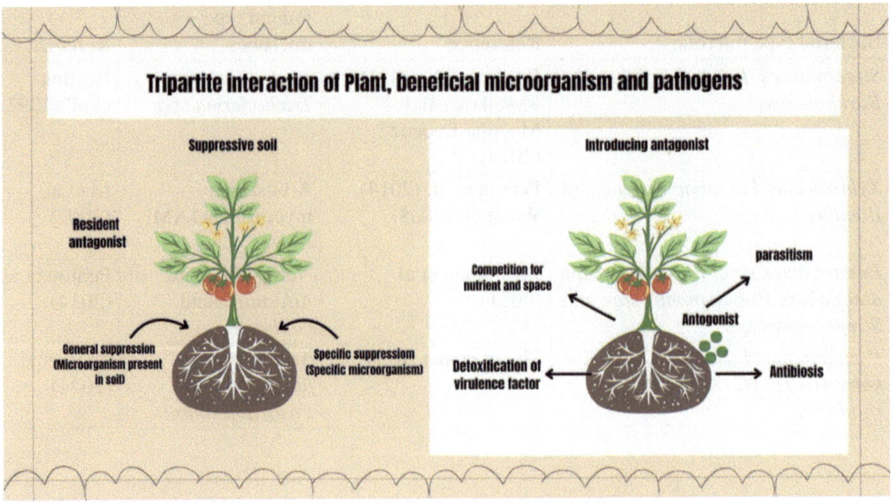

Fig. 10.2 Tritrophic interactions in suppressive soil. (Source: Authors developed)

disease-suppressive soils have been investigated for their ability to combat pathogens such as the oomycetes and fungi including *Pythium ultimum, Pythium irregulare, Pythium aphanidermatum, Phytophthora nicotianae, Phytophthora capsici, Phytophthora cinnamomi, Rhizoctonia solani,* and *Fusarium oxysporum*. Soil suppressiveness to *Fusarium* wilt in tomatoes has been attributed to competition for carbon and iron. In this process, pathogenic isolates of *Fusarium oxysporum* in the rhizosphere compete with the non-pathogenic endophytic strain Fo47 of *F. oxysporum* for carbon, while wild populations of fluorescent pseudomonads compete for iron.

Suppression of *Rhizoctonia* diseases has been found to be triggered by specific groups of microorganisms working together synergistically. The effectiveness of soil amendments with green composts in controlling *Rhizoctonia* damping-off, collar rot, and root rot has proven to be highly variable and unpredictable across various horticultural soilless systems. The genus *Streptomyces* has been recognized as a highly effective biological control agent (BCA) against Fusarium wilt in cucumber and tomato. It is known for its suppressive effects against both bacterial and fungal wilt diseases. The suppression of *Verticillium dahliae* in eggplant, when composted tomato waste is added to conducive soil, is thought to be due either to the biological activity of *Bacillus* and fluorescent pseudomonads or to systemic resistance induced by *Trichoderma* species. Mendes et al. (2011) used a microbiomic approach to study suppressive soil that had been affected by an outbreak of *Rhizoctonia solani.* They discovered that specific members of the Pseudomonadaceae family, which produce lipopeptides encoded by non-ribosomal peptide (NRPS) genes, were crucial to the soil's suppressive ability. However, they found no significant differences in the dominant bacterial groups between disease-suppressive and conducive soils. Studies led by David M. Weller's research team have shown that these suppressive effects are linked to the proliferation of populations of fluorescent *Pseudomonas* spp. capable of producing the compound 2,4-diacetylphloroglucinol (DAPG) (Exposito et al. 2017). Mendes et al. (2011) applied a microbiomic approach to suppressive soil that was induced by the outbreak of disease caused by *R. solani* and revealed that certain members of *Pseudomonadaceae* that produce a lipopeptide encoded by non-ribosomal peptide (NRPS) genes were the key populations in suppressive soil, while no significant variability in dominant bacterial groups was detected between disease suppressive and conducive soils. In suppressive soils against *F. oxysporum*, Alabouvette (1999) suggested that the suppressive effects are likely attributed to non-pathogenic *Fusarium* species. A recent microbiome study by Siegel-Hertz et al. (2018) found that fungal diversity varies between soils that suppress *F. oxysporum* and those that promote its growth. They observed that several genera of fungi and bacteria known for their antagonistic activities against *F. oxysporum* are either exclusively present or more abundant in suppressive soils.

8 Metagenomics

Microbial genome sequencing has become a key tool in applied soil microbiology and ecology because of its growing affordability, faster sequencing capabilities, and enhanced data quality (De Corato 2020). Metagenomic analysis, which aims to identify all microbial genes present in a sample, requires extensive coverage, which can limit the number of samples processed in a single run and increase the cost per sample. Next-generation sequencing (NGS) examines the species and functional diversity of soil microbial communities without the need for culture media through two main methods: (i) amplicon sequencing (or metabarcoding), which involves amplifying and sequencing specific marker gene families, and (ii) metagenomics, which entails random shotgun sequencing of the entire genomic content of microbial communities. It is crucial to distinguish between these two methods, as they are sometimes mistakenly referred to as metagenomics or confused with one another. Authors suggest using the term "metabarcoding" for amplicon-based techniques and reserving "metagenomics" for untargeted shotgun sequencing. Both techniques have effectively identified and studied uncultivable microorganisms without the need for isolating single colonies on growth media.

Metagenomics offers valuable insights into the study of agroecosystems by revealing the extensive microbial diversity present in soil. This approach helps in identifying novel molecules with potential therapeutic uses, advancing biotechnology, and promoting sustainable agricultural practices. Historically, research has predominantly concentrated on the biological characteristics of cultivable microorganisms and their potential to mitigate environmental impacts. While this has achieved some success, it is limited because it overlooks the contributions of the majority of soil microorganisms. Hjort et al. (2009) observed that chitinase genes in Swedish soil known for its suppressive properties against phytopathogens, we compared data from a metagenomic DNA library with direct DNA extraction from the soil. Our analysis revealed that nearly all chitin-degrading isolates belonged to genera known for chitinase production, such as *Streptomyces, Stenotrophomonas, Pseudomonas,* and *Bacillus.* The most prevalent chitinase-producing isolates were identified as *S. mutomycini* and/or *S. clavifer*, which also harbored the most common chitinase gene variant. TRF159, associated with *S. mutomycini* and/or *S. clavifer* 16S rRNA genes, was highly abundant in the soil. These findings collectively indicate that the chitinase and 16S rRNA genes detected through molecular methods likely correspond to the Streptomyces spp. isolates found in the same soil. Consequently, these isolates are probably responsible for the observed chitinase production in the suppressive soil and may hold potential for biocontrol of certain soil-borne fungal diseases. The soil in Sweden harbors bacteria that both inhibit the growth of Rhizoctonia and produce chitinase, an enzyme that breaks down chitin (Adesina et al. 2007).

9 Practical Utility/Case Studies

Additionally, this soil was previously identified as suppressive to clubroot disease caused by *Plasmodiophora.* Since both the cell walls of *Plasmodiophora* spores and the mycelium of *Rhizoctonia* contain chitin, chitinase activity might be an effective means of combating these plant pathogens. However, it's also possible that other abiotic or unknown biological factors contribute to the soil's suppressive properties (Hardeman and Sjoling 2007). The study by Karin Hjort et al. 2014 utilized a functional metagenomics approach to explore chitin-degrading enzymes from an uncultured bacterial community, employing a fosmid library derived from a suppressive soil metagenome. This research led to the discovery of a novel bacterial chitinase named Chi18H8, which exhibits antifungal activity against several significant crop pathogens. Sequence analysis revealed that the chi18H8 gene encodes a 425-amino acid protein with a molecular weight of 46 kDa. This protein includes an N-terminal signal peptide, a catalytic domain with the conserved active site F175, and an additional chitinase insertion domain. Chi18H8 was expressed in *Escherichia coli* using the pGEX-6P-3 vector and subsequently purified. Characterization of the enzyme demonstrated that Chi18H8 primarily functions as a chitobiosidase, showing optimal activity at 35 °C and pH values below 6, indicating its role as an exochitinase on native chitin. Notably, Chi18H8 is the first chitinase isolated in pure form from a metagenome library and holds promise as a potential agent for controlling fungal diseases in crops.

Bacterial chitinases are promising candidates for this purpose because they can degrade chitin, a key component of fungal cell walls. Additionally, chitinolytic enzymes are of broad interest for various biotechnological applications. They hold potential in biofuel production and in bioconversion processes of shellfish waste, which can lead to valuable products like chitosan and chito oligosaccharides for the pharmaceutical industry (Li and Greene 2010). The Chi18H8 enzyme derived from the metagenome exhibits optimal activity and stability at acidic pH levels and mesophilic temperatures, which aligns with findings for family 18 chitinases from cultured microorganisms (Bhattacharya et al. 2007). However, Chi18H8 uniquely demonstrates selective antagonistic activity against three significant plant pathogens *C. gloeosporioides, F. oxysporum* and *F. graminearum* - reflecting the suppressive characteristics of the soil from which it was obtained. Notably, while antifungal activity has been observed in family 19 glycosyl hydrolases, Chi18H8's antifungal properties are a distinct feature among family 18 chitinases.

10 Future Thrust and Conclusions

Over a century of research into suppressive soils has provided critical inputs and insights into disease suppression. The soil's ability to suppress plant pathogens largely stems from its biological activities, which will be better understood thorough microbiome studies. The cases discussed here highlight this point. Clarifying the microbial roles in suppressive soils will undoubtedly pave the way for sustainable agricultural practices in the future. Moreover, future research will provide new foundational insights essential for developing cost-effective and eco-friendly strategies to manage disease suppression. There remains a challenge in utilizing metagenomics to uncover how microbiomes counteract pathogens. Additionally, investigating uncultured microorganisms for producing antibiotics like texiobactin and growth factors such as siderophores shows promise. Moreover, understanding the intricate interactions between abiotic and biotic factors and their variations across different soil systems requires further investigation. This complex framework can be expanded by promoting integrated competencies through a transdisciplinary approach necessary to grasp the complexities of soil systems. This approach aims to identify and interpret suppressive mechanisms, thereby enhancing the practical applications of Disease Suppressive Soils (DSS) in agriculture.

References

Adesina MF, Lembke A, Costa R, Speksnijder A, Smalla K (2007) Screening of bacterial isolates from various European soils for in vitro antagonistic activity towards *Rhizoctonia solani* and *Fusarium oxysporum*: site-dependent composition and diversity revealed. Soil Biol Biochem 39:2818–2828

Alabouvette C (1999) Fusarium wilt suppressive soils: an example of disease suppressive soils. Australas J Plant Pathol 28:57–64

Baker KF, Cook RJ (1974) Biological control of plant pathogens. W.H. Freeman, San Francisco

Bhattacharya D, Nagpure A, Gupta RK (2007) Bacterial chitinases: properties and potential. Crit Rev Biotechnol 27:21–28

Borneman J, Becker JO (2007) Identifying microorganisms involved in specific pathogen suppression in soil. Annu Rev Phytopathol 45:153–172

Brader G, Compant S, Mitter B, Trognitz F, Sessitsch A (2014) Metabolic potential of endophytic bacteria. Curr Opin Biotechnol 27:30–37

Brady NC (1996) The nature and properties of soils, 10th edn. Reprinted in India by Special Arrangement with Prentice Hall, Inv, Englewood Clifffs

Cook RJ (2014a) Plant health management: pathogen suppressive soils. In: Alfen V, Neal K (eds) Encyclopedia of agriculture and food systems. Academic, pp 441–455

Cook RJ (2014b) Plant health management: pathogen suppressive soils. In: Encyclopedia of agriculture and food systems. Elsevier, Amsterdam, pp 441–455

Cook RJ, Baker KF (1983) The nature and practice of biological control of plant pathogens. The American Phytopathological Society, St. Paul, p 539

De Corato U (2020) Soil microbiota manipulation and its role in suppressing soil-borne plant pathogens in organic farming systems under the light of microbiome-assisted strategies. Chem Biol Technol Agric 7:17

Dinca LC, Grenni P, Onet C, Onet A (2022) Fertilization and soil microbial community: a review. Appl Sci 12(3):1198

Duran P, Thiergart T, Garrido-Oter R, Agler M, Kemen E, Schulze-Lefert P, Hacquard S (2018) Microbial interkingdom interactions in roots promote Arabidopsis survival. Cell 175(973–983):e914

Exposito RG, de Bruijn I, Postma J, Raaijmakers JM (2017) Current insights into the role of rhizosphere bacteria in disease suppressive soils. Front Microbiol 8:2529

Garbeva PV, Van Veen JA, Van Elsas JD (2004) Microbial diversity in soil: selection of microbial populations by plant and soil type and implications for disease suppressiveness. Annu Rev Phytopathol 42:243–270

Guerra CA, Heintz-Buschart A, Sikorski J, Chatzinotas A, Guerrero-Ramírez N, Cesarz S, Beaumelle L, Rillig MC, Maestre FT, Delgado-Baquerizo M, Buscot F (2020) Blind spots in global soil biodiversity and ecosystem function research. Nat Commun 11(1):1–13. https://doi.org/10.1038/s41467-020-17688-2

Hadar Y, Papadopoulou KK (2012) Suppressive composts: microbial ecology links between abiotic environments and healthy plants. Annu Rev Phytopathol 50(50):133–153

Hardeman F, Sjoling S (2007) Metagenomic approach for the isolation of a novel low-temperature-active lipase from uncultured bacteria of marine sediment. FEMS Microbiol Ecol 59:524–534

Hjort K, Bergstrom M, Modupe FA, Janet K, Jansson KS, Sjoling S (2009) Chitinase genes revealed and compared in bacterial isolates, DNA extracts and a metagenomic library from a phytopathogen-suppressive soil. FEMS Microbiol Lett 71(2):197–207

Hjort K, Presti I, Elvang A, Marinelli F, Sjoling S (2014) Bacterial chitinase with phytopathogen control capacity from suppressive soil revealed by functional metagenomics. Appl Microbiol Biotechnol 98:2819–2828

Hoitink HAJ, Stone AG, Han DY (1997) Suppression of plant disease by composts. Hortic Sci 32:184–187

Hoorman JJ (2020) Understanding soil microbes and nutrient recycling. Ohio State University Extension, USA

Jayaraman S, Naorem AK, Lal R, Dalal RC, Sinha NK, Patra AK, Chaudhari SK (2021) Disease-suppressive soils – beyond food production: a critical review. J Soil Sci Plant Nutr 21:1437–1465

Kwak YS, Weller DM (2013) Take-all of wheat and natural disease suppression: a review. Plant Pathol J 29(2):125–135

Kyselková M, Moënne-Loccoz Y (2012) *Pseudomonas* and other microbes in disease-suppressive soils. In: Organic fertilisation, soil quality and human health. Springer Netherlands, pp 93–140

Kyselkova M, Kopecky J, Frapolli M, Defago G, Sagova-Mareckova M, Grundmann GL (2009) Comparison of rhizobacterial community composition in soil suppressive or conducive to tobacco black root rot disease. ISME J.3:1127–1138

Li H, Greene LH (2010) Sequence and structural analysis of the chitinase insertion domain reveals two conserved motifs involved in chitin binding. PLoS One 5:e8654

Li B, Ravnskov S, Xie G, Larsen J (2007) Biocontrol of *Pythium* damping-off in cucumber by arbuscular mycorrhiza – associated bacteria from the genus *Paenibacillus*. BioControl 52:863–875

Mendes R, Kruijt M, de Bruijn I, Dekkers E, van der Voort M, Schneider JHM, Piceno YM, DeSantis TZ, Andersen GL, Bakker PAHM, Raaijmakers JM (2011) Deciphering the rhizosphere microbiome for disease-suppressive bacteria. Science 332:1097–1100

Ng TFF, Dufy S, Polston JE, Bixby E, Vallad GE, Breitbar M (2011) Exploring the diversity of plant DNA viruses and their satellites using vector-enabled metagenomics on whiteflies. PLoS One 6(4):e19050

Olatinwo R, Yin B, Becker JO, Borneman J (2006) Suppression of the plant-parasitic nematode *Heterodera schachtii* by the fungus *Dactylella oviparasitica*. Phytopathology 96:111–114

Penton CR, Gupta VVSR, Tiedje JM, Neate SM, Ophel-Keller K, Gillings M, Roget DK (2014) Fungal community structure in disease suppressive soils assessed by 28S LSU gene sequencing. PLoS One 9(4):e93893

Scholthof KBG (2007) The disease triangle: pathogens, the environment and society. Nat Rev Microbiol 5(2):152–156

Siegel-Hertz K, Edel-Hermann V, Chapelle E, Terrat S, Raaijmakers JM, Steinberg C (2018) Comparative microbiome analysis of a Fusarium wilt suppressive soil and a Fusarium wilt conducive soil from the Chateaurenard region. Front Microbiol 9:568

Thilagavathi R, Nakkeeran S, Balachandar D, Raguchander T, Samiyappan R (2021) Bacterial community analysis on Sclerotium-suppressive soil. Arch Microbiol 203:4539–4548

Van Elsas JD, Trevors JT (1997) Modern soil microbiology. Marcel Dekker, New York

Weller DM, Raaaijmakers JM, MacSpadden Gardener BB, Thomashow LS (2002) Microbial populations responsible for specific soil suppressiveness to plant pathogens. Annu Rev Phytopathol 40:309–348

Weller DM, Landa BB, Mavrodi OV, Schroeder KL, De La Fuente L, Blouin Bankhead S, Allende Molar R, Bonsall RF, Mavrodi DV, Thomashow LS (2007) Role of 2,4-diacetylphloroglucinol-producing fluorescent *Pseudomonas* spp. in the defense of plant roots. Plant Biol 9:4–20

Wong PTW, Baker R (1984) Suppression of wheat take-all and *Ophiobolus* patch by fluorescent pseudomonads from a fusarium-suppressive soil. Soil Biol Biochem 16(4):397–403

Wu T, Chellemi DO, Graham JH, Martin KJ, Rosskopf EN (2008) Comparison of soil bacterial communities under diverse agricultural land management and crop production practices. Microbial Ecol 55(2):293–310

Yuan X, Hong S, Xiong W, Raza W, Shen Z, Wang B, Dini-Andreote F (2021) Development of fungal-mediated soil suppressiveness against Fusarium wilt disease via plant residue manipulation. Microbiome 9:200

Zheng Y, Han X, Zhao D, Wei K, Yuan Y, Li Y, Zhang CS (2021) Exploring biocontrol agents from microbial keystone taxa associated to suppressive soil: a new attempt for a biocontrol strategy. Front Plant Sci 12:655673

Index

S. Gnanamanickam, M. Höfte (eds.), *Disease Suppressive Tropical Soils*, https://doi.org/10.1007/978-3-032-08100-1

D

MIX
Papier aus verantwortungsvollen Quellen
Paper from responsible sources
FSC® C105338

If you have any concerns about our products,
you can contact us on
ProductSafety@springernature.com

In case Publisher is established outside the EU,
the EU authorized representative is:
Springer Nature Customer Service Center GmbH
Europaplatz 3, 69115 Heidelberg, Germany

Printed by Libri Plureos GmbH
in Hamburg, Germany